MSACL 2022

The Association for
Mass Spectrometry &
Advances in the Clinical Lab

The 12th Annual Conference
of
The Association for

Mass Spectrometry &
Advances in the Clinical Lab

Monterey, California

April 5 - 8, 2022

Monterey Conference Center

The Association is a non-profit 501(c)(3) tax-exempt California Corporation with the mission of furthering education in advanced technologies for use in the clinical laboratory.

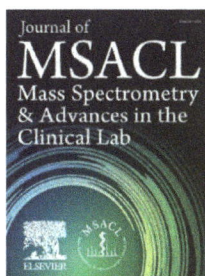

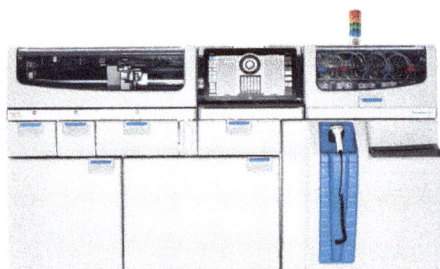

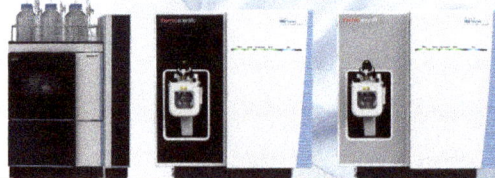

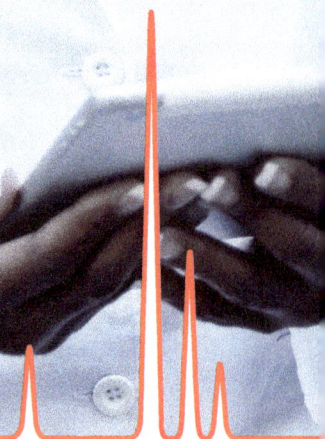

Table of Contents

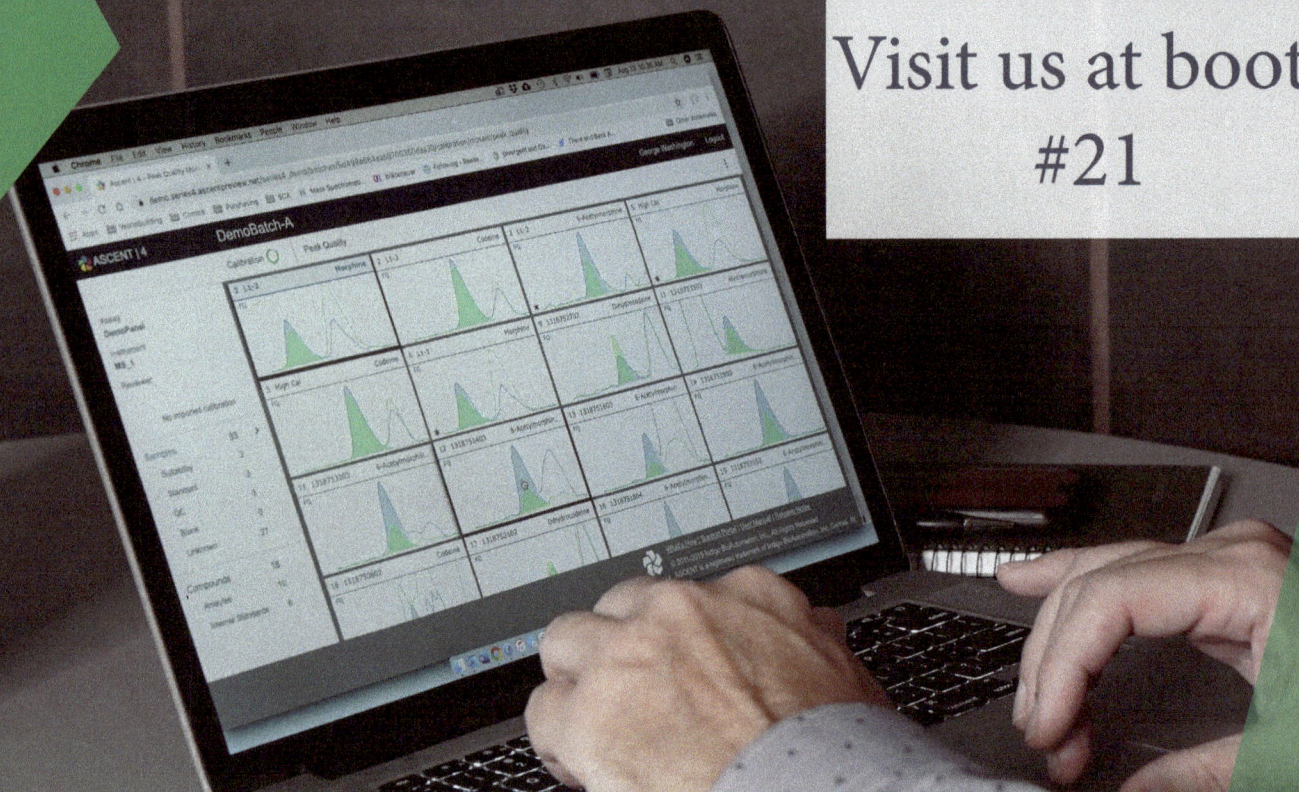

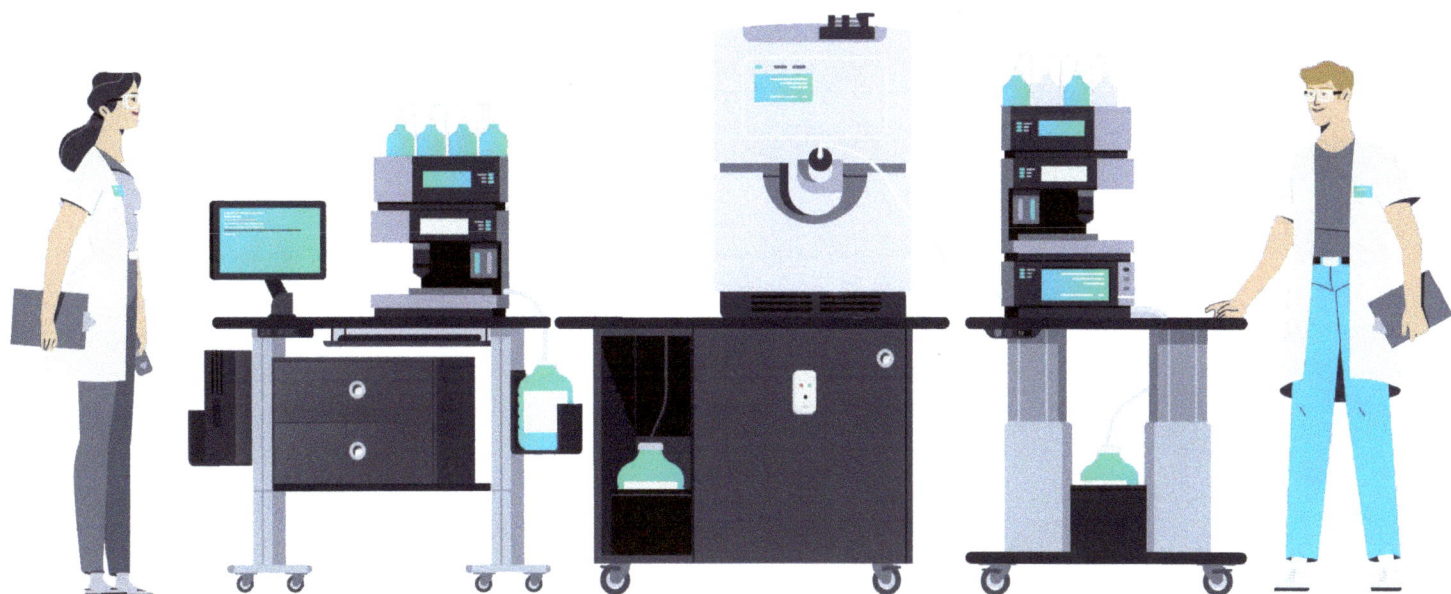

Learn what the latest innovations in mass spectrometry can do for your lab!

We're thrilled to be back at MSACL and see you all in person once again!

Be sure to come by our booth, discuss the latest advancements for analysis in clinical and research labs and pick up some fun giveaways!

Visit
SCIEX at
Booth #1

Don't miss these posters

#13a | Wednesday | 11:00 a.m.
Highly sensitive quantification of proteins from the SARS-CoV-2 antigen in nasopharyngeal swab samples

#3b | Wednesday | 1:00 p.m.
Highly sensitive MS/MS detection for confident identification of potent novel synthetic opioids and their metabolites

#9b | Wednesday | 1:00 p.m.
Rapid LC-MS/MS method for monitoring bio-relevant levels of per- and polyfluoroalkyl substances (PFAS) in serum

#14a | Thursday | 11:00 a.m.
Selective separation and sensitive detection of intact parathyroid hormone and variants by CESI-MS/MS

#12b | Thursday | 1:00 p.m.
Analysis of estrogens in plasma with rapid chromatography and reduced sample volume

#14b | Thursday | 1:00 p.m.
Improved sensitivity for aldosterone using the unique MRM3 quantification workflow

You can request this complete poster pack now – just scan the QR code and submit the form:

SCIEX
The Power of Precision

Steering Committee

Cory Bystrom, PhD
Ultragenyx

Chair

Daniel Holmes, MD, FRCPC
St. Paul's Hospital

Outgoing Chair

In alphabetical order ...

Michael Angelo, MD, PhD
Stanford University School of Medicine

Timothy Collier, PhD
Quest Diagnostics

David Herold, MD, PhD
MSACL, University of California San Diego and VA San Diego Medical Center

Kara Lynch, PhD
University of California San Francisco

Laura Sanchez, PhD
University of California, Santa Cruz

Stefani Thomas, PhD, DABCC, NRCC
University of Minnesota

Practical Training Committee

Grace van der Gugten
Government of Alberta, Medical Examiner's Office

Chair and Global Coordinator

Deborah French, PhD
UCSF

Jacqueline Hubbard, PhD, DABCC
QualiTox Laboratoires

Conference Overview : Monday

	Monday
Time	**Sessions**
1600 1900	**Early Badge Pickup** *Location: De Anza Foyer*
1600 2000	**Welcome Hospitality Reception – Drinks and very light snacks, not intended as Dinner substitute** *Location: Jacks*

Places to Eat

The Monterey Conference Center (MCC) is within 2 blocks of the ocean.

Immediately proximate to the MCC, going away from the ocean, is Alvarado St, which hosts a plethora of restaurants and shops, including Alta Bakery, Crepes of Brittany, Walgreens Pharmacy and Trader Joe's grocery store.

Towards the ocean there are two wharfs, one is the touristy Old Fisherman's Wharf with several restaurants, and the second is the commercial wharf, which has the Sandbar Grill, with a smaller, more local, feel.

British Pubs and Breweries
Alvarado Brewery, Dustbowl Brewing, London Bridge Pub, The Crown and Anchor, Britannia Arms, Peter B's (part of the Portola).

The Alvarado Brewing company is a standout brewery. We have not yet been to Dustbowl, so, at own risk. Alvarado houses a restaurant while the Dustbowl hosts a food truck.

Cibo is an Italian Restaurant, on the near corner of Alvarado, that doubles as the local late night bar - live music - dancing venue. It has $5 drinks all night on Thursday.

Cannery Row includes **A Taste of Monterey Wine Market and Bistro** where you can sample all the local wines.

Some upscale options

1. Cafe Fina
2. Stokes Adobe (Closed Wednesday)
3. Estéban Restaurant
4. Passionfish (Pacific Grove)
5. Vesuvio (Carmel by the Sea)

Conference Overview : Tuesday, Wednesday

Tuesday

Time	Sessions
630	**Registration Desk Opens** - *De Anza Foyer*
800 1600	**Vendor Booth Set-Up** *Location: Exhibit Hall - Serra*
800 1200	**Workshop: A Clinical Proteomics Primer** *Location: De Anza 1*
800 1200	**Workshop: Why We Fail at Biomarkers** *Location: De Anza 3*
800 1200	**Short Course: Data Science 101** *Location: Bonsai*
800 950	**Short Course: LC-MSMS 101 : Hands-On Training** *Location: Colton*
900 1200	**Workshop : How to Convince Admin THEY Want to Buy You a Mass Spectrometer** *Location: De Anza 2*
1000 1150	**Short Course: LC-MSMS 101: Hands-On Training** *Location: Colton*
1000 1145	**Workshop: Design of Experiments for Development and Optimization of LC-MS Clinical Diagnostic Assay** *Location: Steinbeck 1*
1200 1400	**Short Course & Workshop Lunch Mixer** *Location: Steinbeck Foyer*
1215 1345	**Workshop: Ion Mobility in the Clinical Lab?** *Location: Steinbeck 1*
1400 1425	**Welcome Orientation** *Location: De Anza*
1425 1445	**State of the Science** *Location: De Anza*
1445	**Break -** *De Anza Foyer*
1500 1545	**Beyond the Human Genome: A Million Person Precision Population Health Project (Hood)** *Location: De Anza*
1545	**Break -** *De Anza Foyer*
1600 1645	**The Clinical Laboratory Perspective on Wellness Testing: Let's Take a Look Under the Hood (Baird)** *Location: De Anza*
1645	**Break -** *De Anza Foyer*
1700 1800	**The Michael S. Bereman Award for Innovative Clinical Proteomics : Taking a Step Back to Move Proteomics Forward in the Clinical Lab (DiMarco)** *Location: De Anza*
1800 2100	**Opening Exhibits Reception** *Location: Exhibit Hall - Serra*
1900 2000	**Mentor Booth Tours** *Location: Exhibit Hall – Serra (meet at poster #50)*
2000 2100	**Troubleshooting Posters (p 63-64)** *Location: Exhibit Hall – Serra (meet at poster #28a)*
2100 2200	**Hospitality Lounge** *Location: Jacks (Portola)*

Wednesday

Time	Sessions
630	**Registration Desk Opens -** *De Anza Foyer*
700 745	**Round Table Interest Groups** *Location: De Anza Foyer*
800 855	**Glycoproteins as Biomarkers for Cancers (Lebrilla)** *Location: De Anza*
900 930	**How Can Proteomics Fulfill the Unmet Needs of Effective Drug Treatment Stratification for Patients with Ovarian Cancer? (Thomas)** *Location: De Anza*
930 945	**Coffee Break** *Location: Exhibit Hall - Serra*
945 1015	**N-linked Glycans in Human Disease: From New Tools to Translational and Preclinical Studies (Angel)** *Location: De Anza*
1015 1100	**Panel Discussion** *Location: De Anza*
1100 1200	**Poster Session** *Location: Exhibit Hall - Serra*
1200 1300	**Industry Workshops – Pre-registration Required** *Location: De Anza 1-3*
1300 1400	**Poster Session** *Location: Exhibit Hall - Serra*
1400 1500	**Scientific Session 1** *Location: De Anza 1-3, Bonsai*
1500 1515	**Coffee Break** *Location: Exhibit Hall – Serra*
1515 1615	**Scientific Session 2** *Location: De Anza 1-3, Bonsai*
1615 1630	**Coffee Break** *Location: Exhibit Hall – Serra*
1630 1730	**Scientific Session 3** *Location: De Anza 1-3, Bonsai*
1730 1830	**Happy Hour in Exhibit Hall** *Location: Exhibit Hall - Serra*
1745 1815	**Troubleshooting Posters (p 63-64)** *Location: Exhibit Hall – Serra (meet at poster #29a)*
1830 1930	**Hospitality Reception** Pre-dinner drinks to meet-up with others and then head out to dinner on own. *Location: Jacks (Portola)*
1830	**Dinner on Own** *Location: Off-site*
1845 2200	**FeMS Networking Event** *Separate Registration Required* *Visit Registration Desk if Interested* *Location: Ferrantes (Marriot)*

Conference Overview : Thursday, Friday

Thursday

Time	Sessions
630	**Registration Desk Opens** - *De Anza Foyer*
700 745	**Round Table Interest Groups** *Location: De Anza Foyer*
800 855	**Proteins in Space: Statistical Approaches to Understand the Spatial Organization and Structure of Proteins in Complex Tissues (Englehardt)** *Location: De Anza*
855 900	**Analytik Jena Industry Brief (5m)** *Location: De Anza*
900 915	**Coffee Break** *Location: Exhibit Hall - Serra*
915 1000	**Statistical Considerations for Biomarker Discovery Experiments: From a Model Organism to Clinical Study (Choi)** *Location: De Anza*
1000 1015	**Coffee Break** *Location: Exhibit Hall - Serra*
1015 1100	**Adapting to the New Normal: Navigating the Rollercoaster of SARS-CoV-2 Testing Needs While Building Long Term Capabilities (Mathias)** *Location: De Anza*
1100 1200	**Poster Session** *Location: Exhibit Hall - Serra*
1200 1300	**Industry Workshops – Pre-registration Required** *Location: De Anza 1-3*
1300 1400	**Poster Session** *Location: Exhibit Hall - Serra*
1400	**Exhibits Closed** *Location: Exhibit Hall - Serra*
1400 1500	**Scientific Session 4** *Location: De Anza 1-3, Bonsai*
1500 1515	**Break** *Location: De Anza Foyer*
1515 1615	**Scientific Session 5** *Location: De Anza 1-3, Bonsai*
1615 1630	**Break** *Location: De Anza Foyer*
1630 1730	**Scientific Session 6** *Location: De Anza 1-3, Bonsai*
1745 2000	**Dinner & British-Style Trivia Night with Tim** *Location: San Carlos Ballroom in Marriott*
2000	**After Hours Suggestion** *Location: Alvarado Beer Garden, Cibo ($5 drinks all night)*

Friday

Time	Sessions
700 800	**Monterey Challenge Run/Walk** *Location: Meet at De Anza Foyer*
800	**Registration Desk Opens** - *De Anza Foyer*
900 1000	**Addressing Hurdles in Clinical Translation of Targeted Proteomics (Whiteaker)** *Location: De Anza*
1000 1030	**Newborn Screening by Mass Spec Meets Newborn Screening by DNA Sequencing (Gelb)** *Location: De Anza*
1030 1050	**Coffee Break** *Location: De Anza Foyer*
1050 1120	**Utilization of Mass Spectrometry to Discover and Develop Novel Biomarkers to Support Drug Development (Anania)** *Location: De Anza*
1120 1155	**Panel Discussion** *Location: De Anza*
1155 1200	**Closing Statements** *Location: De Anza*
1200 1400	**Boxed Lunch Pick-Up and Mixer** *Location: Jacks (Portola)*
1215 1345	**Workshop: Rethinking the Traditional Workflow for Urine Toxicology Testing** *Location: Bonsai*
1400 1700	**Workshop: Pre-analytical considerations as prerequisite for successful clinical application of lipidomics** *Location: De Anza 1*
1400 1800	**Short Course: LC-MSMS 201 : Understanding and Optimization of LC-MS/MS to Develop Successful Methods for Identification and Quantitation in Complex Matrices** *Location: De Anza 2*
1400 1800	**Workshop: CLSI C64 - Supporting development of quantitative protein and peptide assays for clinical use** *Location: De Anza 3*
1400 1800	**Short Course: Sample Prep 201 : Sample Preparation and Alternative Matrices for LC-MS Assays** *Location: Bonsai*
1400 1800	**Short Course: Data Science 201 : Going Further With R: Tackling Clinical Laboratory Data Manipulation and Modeling** *Location: Colton*
1800 2000	**Closing Dinner Reception** *Location: Jacks (Portola)*
2000	**After Hours Suggestion** *Location: Peter B's, Cibo*

Top-Tier Sponsorship & Educational Grant Support

Platirum — $17,000

GoldenWest Diagnostics, LLC™

SCIEX

ThermoFisher SCIENTIFIC

Gold — $12,000

indigo bioAutomation

Educational Grants — $10,000

SCIEX

Waters THE SCIENCE OF WHAT'S POSSIBLE.™

& Brian Kelly

Workshops : Tuesday

These workshops do not require separate registration to attend.
They are included with your MSACL 2022 Registration.

Workshop: A Clinical Proteomics Primer
Tuesday 8:00-12:00 @ De Anza 1

ANDY HOOFNAGLE, MD, PhD
UNIVERSITY OF WASHINGTON

CHRISTOPHER SHUFORD, PH.D.
LABCORP

Syllabus
1. Protein vs Peptide Measurands
2. Workflows
3. Sample Preparation (Digestion & Enrichment)
4. Internal standards
5. Calibration
6. Validation
7. Quality control

Workshop: Why We Fail at Biomarkers
Tuesday 8:00-12:00 @ De Anza 3

TIM GARRETT, PhD
UNIVERSITY OF FLORIDA COLLEGE OF MEDICINE

Syllabus
1. Metabolomics in clinical research
2. Quality control for better method assessment
3. Experimental design
4. Statistical analyses with validation
5. Real-world samples

Workshop : How to Convince Admin THEY Want to Buy You a Mass Spectrometer
Tuesday 9:00-12:00 @ De Anza 2

JOSHUA HAYDEN, PHD, DABCC, FACB
NORTON HEALTHCARE

JUAN DAVID GARCIA, MBA MT
UNIVERSITY OF TEXAS MEDICAL BRANCH

Syllabus
1. How to assemble the worst business case ever (and guarantee failure)
2. From LCMS and qTOF to ROI and DEPR (speaking the language of finance)
3. Estimating costs
4. Estimating reimbursements
5. What's wrong with my business case?

Workshop: Design of Experiments for Development and Optimization of LC-MS Clinical Diagnostic Assay
Tuesday 10:00-11:45 @ Steinbeck 1

MARGRET THORSTEINSDOTTIR, PHD
FACULTY OF PHARMACEUTICAL SCIENCES, UNIVERSITY OF ICELAND

FINNUR EIRIKSSON, PHD
ARTICMASS

Syllabus
1. Design of Experiments (DoE) – Get it right from the beginning
2. Basic concept and assessment of DoE
3. Optimization of LC-MS based clinical assay by DoE

Workshop: Ion Mobility in the Clinical Lab?
Tuesday 12:15-13:45 @ Steinbeck 1

Pick-up a ==boxed lunch== from the Steinbeck foyer & then enjoy this session (included w MSACL registration).

CHRISTOPHER CHOUINARD, PhD
FLORIDA INSTITUTE OF TECHNOLOGY

ROBIN KEMPERMAN, PhD
CHILDREN'S HOSPITAL OF PHILADELPHIA

Syllabus
1. Basic Operating Conditions of IMS: Electric field application, experimental conditions (temperature, pressure, gas composition)
2. Different IMS techniques: Drift tube/traveling wave, field asymmetric/differential mobility, emerging techniques (i.e., TIMS, SLIM, cIMS, etc.)
3. Applications: Current examples from metabolomics, lipidomics, and proteomics

Workshops : Friday

Workshop: Rethinking the Traditional Workflow for Urine Toxicology Testing
Friday 12:15-13:45 @ Bonsai

Pick-up a ==boxed lunch== from Jacks and then enjoy this session (included w MSACL registration).

MELISSA BUDELIER, PhD
TRICORE REFERENCE LABORATORIES

BENJAMIN BEPPLER
TRICORE REFERENCE LABORATORIES

Syllabus
1. Discuss several alternatives for urine drug testing, such as direct to mass spectrometry testing and 'hybrid' panels.
2. Discuss considerations for selecting the appropriate analytes for urine drug testing panels.
3. Discuss potential pitfalls associated with certain classes of drugs.
4. Introduce our recent efforts to implement the use of interpretive reporting for pain management clients.

Workshop: Pre-Analytical Considerations as Prerequisite for Successful Clinical Application of Lipidomics

Friday 14:00-17:00 @ De Anza 1

ROBERT GURKE, PhD
FRAUNHOFER INSTITUTE FOR TRANSLATIONAL MEDICINE AND PHARMACOLOGY ITMP

BO BURLA, PhD
SLING @ NATIONAL UNIVERSITY OF SINGAPORE

MARGRET THORSTEINSDOTTIR, PhD
FACULTY OF PHARMACEUTICAL SCIENCES, UNIVERSITY OF ICELAND

ANNE K. BENDT, PhD
SINGAPORE LIPIDOMICS INCUBATOR (SLING), NATIONAL UNIVERSITY OF SINGAPORE

Syllabus
1. Pre-Analytical factors influencing lipid concentrations
2. Capillary vs. venous blood sampling – the potential of microsampling
3. Cohort & Study design for lipidomics in clinical research

Workshop: CLSI C64 – Supporting Development of Quantitative Protein and Peptide Assays for Clinical Use

Friday 14:00-18:00 @ De Anza 3

CORY BYSTROM, PhD
MSACL 2022 CHAIR,
ULTRAGENYX

RUSSELL GRANT, PhD
LABCORP

Syllabus
1. Introduction to C64, philosophy and scope
2. Interactive discussion of each chapter

Short Courses : Tuesday

Short courses are composed of Segments that occur over multiple days, which are mostly online. You can attend the in-person segment at MSACL to preview any course without additional registration*, however, to join the online Segments you will need to register for each course.

*The LC-MSMS 101 Hands-On Segment, **does** require a separate registration. If you are not registered and want to attend, visit the Registration Desk to inquire if space is available.

Short Course: LC-MSMS 101 : Hands-On Training Session
Tuesday 8:00 – 9:50 & 10:00 – 11:50 @ Colton

Sponsored in part by:

JUDY STONE, MT (ASCP), PHD, DABCC
UCSF

JACQUELINE HUBBARD, PHD, DABCC
QUALITOX LABORATOIRES

ADINA BADEA, PHD, DABCC
LIFESPAN HEALTH/RHODE ISLAND HOSPITAL & THE WARREN ALPERT MEDICAL SCHOOL OF BROWN UNIVERSITY

ROBERT FITZGERALD, PHD, DABCC
UNIVERSITY OF CALIFORNIA SAN DIEGO

** *Supplemental Hands-On Segment In-Person : Main Course was Online* ** *This is the supplemental bonus segment (2 hr) of a 16 hour online short course that took place on March 11-14, 2022. Attendance at this in-person segment is free, but REQUIRES pre-registration (separate from conference registration) with priority given to online course registrants. There will be two instances of this segment (Group 1 and Group 2, both the same), with each to be capped at 20 participants.*

Format and Content

The first 50 min session will include brief instructor demonstrations and then ample hands-on time for attendees to practice troubleshooting tasks, such as
- cutting (and recutting) PEEK tubing correctly
- connecting (and reconnecting) PEEK fittings to LC columns, other components
- changing LC pump check valves
- changing LC injection valve rotor seals
- reviewing chromatography problems caused by leaks, tubing/fitting mistakes and damage, excess LC dead volume, and aged LC components

The second 50 min session is a discussion of real world instrument troubleshooting cases from the instructors' laboratories. Aside from the examples presented, the goal is to develop a standardized approach to troubleshooting

complex LC-MS systems, including
- Know your LC flow path and LC components, how to avoid damaging the MS/MS
- How to look for leaks and sources of overpressure
- Using chromatogram overlays, pressure traces, maintenance chart review, system suitability testing and MS/MS infusion to locate the problem within the instrument

Short Course: Data Science 101 : Breaking up with Excel: An Introduction to the R Statistical Programming Language
Tuesday 8:00-12:00 @ Bonsai

DANIEL HOLMES, MD, FRCPC
ST. PAUL'S HOSPITAL

DUSTIN BUNCH, PHD, DABCC
NATIONWIDE CHILDREN'S HOSPITAL

** *Part In-Person* (optional, also available *pre-recorded*) and **Part** *Online* ** *This is the first segment (4 hr) of a three segment (16hr total), part in-person (optional) and part online, short course. Segment 1 will be available both IN-PERSON on April 5 at the MSACL 2022 conference in Monterey, CA and ONLINE (pre-recorded) if you can't make it in person. Registration for Segment 1 is free (although to attend on-site you must be registered for MSACL 2022). Segments 2 and 3 will take place ONLINE on April 29-30, 2022.*

While the first SEGMENT is FREE, SEGEMENTS 2 and 3 that occur only ***ONLINE*** *are* ***fee-based****. You can register by following the QR code above.*

The course will cover:
1. The major types of R variables: vectors (numerical, character, logical), matrices, data frames and lists.
2. The important classes: numeric, character, list and changing between them
3. Importing data from Excel
4. Dealing with non-numeric instrument data
5. Manipulating and cleansing your data
6. Exporting data to Excel-like format.
7. Basics of tidyverse: dplyr, filter, mutate, join
8. Regressions: ordinary least squares,Passing Bablok, Deming, weighted regressions.
9. Non-linear regressions
10. Looping: Doing things repeatedly
11. group_by and summarize
12. Writing your own functions
13. Making highly customized figures with base plot or ggplot
14. Putting it all together projects:
15. Preparing method comparison regression and Bland Altman plots

    Preparing mass spectrometry data for upload to LIS.

Short Courses : Friday

Short Course: Sample Prep 201 : Sample Preparation and Alternative Matrices for LC-MS Assays
Friday 14:00 – 18:00 @ Bonsai

WILLIAM CLARKE, MBA, PhD
JOHNS HOPKINS UNIVERSITY SCHOOL OF MEDICINE

MARK MARZINKE, PhD
THE JOHNS HOPKINS HOSPITAL

Part In-Person and Part Online ** This in-person activity is Segment 2 (4 hr) of a 3 segment (12 hour total), part in-person and part online, short course. Segments 1 and 3 (took) will take place ONLINE on March 4 & April 22, 2022. The first segment is before the conference, the third segment is after the conference. While attending this IN-PERSON segment is FREE, the **ONLINE** attendance is ***fee-based**.*

This course highlights not only the importance of sample processing in the clinical laboratory environment, but also illustrates the "fit for purpose" application of processing techniques in clinical mass spectrometry. This course discusses the theory behind different specimen preparation methods, strengths and weaknesses of each approach, as well as opportunities for automation.

The first 4 hour online segment covered workshop ground rules, introduction, pain points of LC-MS, specimen processing (tube types, management, etc.), and matrix effects.

The second 4 hour in-person segment will cover dilution and protein precipitation, solid phase extraction, supported liquid extraction, liquid-liquid extraction, and affinity-based sample preparation

The third 4 hour online segment (online) will elaborate on the foundations established in the first two segments, and expand into newer technologies and automated alternatives for sample processing.

Course Syllabus

1. Pain points in clinical LC-MS (*covered on March 4* **online**)
2. Overview of specimen processing in laboratory medicine (*covered on March 4* **online**)
3. Off-line sample processing
4. On-line sample processing
5. Analysis of blood and urine
6. LC-MS of tissue specimens
7. Alternate body fluid specimens (e.g. CSF, breast milk, etc.)
8. Dried specimens as matrices
9. Automation of sample processing

Topics will be covered through lecture, Q&A, Case Studies, and small group exercises.

Short Course: Data Science 201 : Going Further With R: Tackling Clinical Laboratory Data Manipulation and Modeling

Friday 14:00-18:00 @ Colton

PATRICK MATHIAS, MD, PHD
UNIVERSITY OF WASHINGTON

SHANNON HAYMOND, PHD
NORTHWESTERN UNIVERSITY FEINBERG SCHOOL OF MEDICINE

** Part In-Person and Part Online ** *This is the first segment (4 hr) of a 16 hour, part in-person and part online, short course. Segments 2, 3 and 4 will take place ONLINE on April 21,22 and 23, 2022. While attending this IN-PERSON segment is FREE, **ONLINE** attendance is **fee-based**. You can find info to register by following the QR code above.*

Having completed your first steps into the wonderful world of data analysis with R (Data Science 101 with Daniel Holmes), would you like to go further? You've learned the basics of R, so now it's time to put that knowledge to work and tackle some interesting clinical applications. Along the way you will also be introduced to even more of capabilities of R and the tools developed by the amazing R community.

The course will be run over two days and time will be split between lecture sessions, individual problem solving, and a highly interactive group-level data mining of real data sets (there may even be prizes). Like the introductory course, this class will maintain the "no student left behind policy". Students will be given time to solve problems taken from real life laboratory work and to do some more advanced analysis on large scale data sets. All attendees will need to bring a laptop with the R language installed and R Studio interface installed. Students may use Windows, Mac OSX or Linux environments. Both R and R studio are free (as in "Free Beer") and open-source.

Students should be prepared continue to expand their skill in programming – which, as you learned in the introductory course can be a little frustrating, but not as frustrating as not being able to get the computer to do what you want at all!

Course Syllabus

1. Core concepts in reproducible data analysis
2. Introduction to version control
3. Using R Markdown for reproducible reports
4. Advanced file reading capabilities
5. Scaling up your data transformation skills
6. Cleaning dirty data and managing timestamps
7. Joining data sets together
8. Connecting R to databases
9. Prediction with linear regression and classification with logistic regression

Short Course: LC-MSMS 201 : Understanding and Optimization of LC-MS/MS to Develop Successful Methods for Identification and Quantitation in Complex Matrices

Friday 14:00-18:00 @ De Anza 2

ROBERT VOYKSNER, PhD
LCMS LIMITED

** Part In-Person and Part Online ** *This is the first segment (4 hr) of a 16 hour, part in-person and part online, short course. Segments 2, 3 and 4 will take place ONLINE on April 22-24, 2022. While attending this IN-PERSON segment is FREE, the ONLINE attendance is* **fee-based***. You can register by following the QR code above.*

This course is designed for the chromatographer / mass spectrometrist who want to be successful in developing methods, method optimization and solving problems using LC/MS/MS. The course covers the atmospheric pressure ionization (API) techniques of electrospray, pneumatically assisted electrospray and atmospheric pressure chemical ionization (APCI) and atmospheric pressure photo ionization (APPI) using single quadrupole, triple quadrupole, time-of-flight and ion trap mass analyzers.

Discussions of sample preparation and chromatography will target method development and optimization for the analysis of "real-world" samples by LC/MS/MS.

The course highlights the following topics with respect to optimization a method to achieve the best sensitivity, specificity and sample throughput:

1. Optimization ionization in API techniques,
2. understanding and minimizing matrix suppression,
3. relative merits of various LC column lengths, particle sizes and column diameters for LC/MS/MS analysis,
4. introduction into the interpretation of MS/MS spectra,
5. important issues in LC/MS/MS quantitation, and
6. optimization of an quantitative analysis.

Applications of LC/MS/MS to analyze compounds of clinical interest in biological matrices will be discussed throughout the course to emphasize the topics covered.

Invited Speakers

Beyond the Human Genome: A Million Person Precision Population Health Project
Tuesday 15:00-15:45 @ De Anza

LEROY HOOD, MD, PHD

INSTITUTE FOR SYSTEMS BIOLOGY

The vision of this project is that we will develop the infrastructure to employ a data-driven approach to optimizing the health trajectory of individuals for body and brain. We have two large populations (5,000 and 10,000) that have validated this approach for body and brain health, respectively. These studies have led to us pioneering the science of wellness and prevention. This project will require the acquisition of key partners for execution, which will be delineated. We are approaching the Federal Government for funding, as we did for the first Human Genome Project. This project will lead to striking new knowledge about medicine, it will catalyze the initiation of start-up companies and it will catalyze a paradigm shift in healthcare from a disease orientation to a wellness and prevention orientation. This will catalyze the largest paradigm shift in medicine, ever.

Moderated by:

DANIEL HOLMES, MD, FRCPC
ST. PAUL'S HOSPITAL

The Clinical Laboratory Perspective on Wellness Testing: Let's Take a Look Under the Hood
Tuesday 16:00-16:45 @ De Anza

GEOFF BAIRD, MD, PHD

UNIVERSITY OF WASHINGTON

As medical science continues to make gains in the elucidation of disease pathophysiology and the discovery of cures , some have questioned the value of dedicating dwindling financial resources to maintaining wellness rather than to fighting disease per se. While both approaches are meritorious and complementary, neither approach is alone sufficient to ensure the health of a population. One major problem with the focus on wellness is the Bayesian dilemma that the positive predictive value of clinical laboratory testing in apparently healthy people is often low, as the specificities of few clinical tests are high enough to ensure that most positive results are true. The impact of this dilemma on laboratory-based wellness approaches will be discussed.

Moderated by: **DANIEL HOLMES, MD, FRCPC**

The Michael S. Bereman Award for Innovative Clinical Proteomics : Seeing the Forest for the Trees: Taking a Step Back to Move Proteomics Forward in the Clinical Lab

Tuesday 17:00-18:00 @ De Anza

MARI DEMARCO, PhD, DABCC, FACB, FCACB

UNIVERSITY OF BRITISH COLUMBIA

Want to run a new test in your clinical lab that takes multiple days to prep, has a complicated (and costly) calibration scheme, and a detection approach so selective it could miss the analyte of interest? If that doesn't sound appealing, you would be in the majority! While the analytical advantages of mass spectrometry resulted in it decisively displacing ligand binding methods as the gold standard approach for protein quantitation, making progress on the routine testing front has taken additional effort. Here we look at how re-evaluating the status quo in clinical proteomics has helped us take leaps forward and implement protein mass spectrometry to improve patient care.

Moderated by:

CHRISTOPHER SHUFORD, PH.D.
LABCORP

ANDY HOOFNAGLE, MD, PhD
UNIVERSITY OF WASHINGTON

Glycoproteins as Biomarkers for Cancers

Wednesday 8:00-8:55 @ De Anza

CARLITO LEBRILLA, PhD

UC DAVIS

The path from fundamental discoveries with MS to translation and commercialization.

Moderated by:

LAURA SANCHEZ, PhD
UNIVERSITY OF CALIFORNIA, SANTA CRUZ

How Can Proteomics Fulfill the Unmet Needs of Effective Drug Treatment Stratification for Patients with Ovarian Cancer?

Wednesday 9:00-9:30 @ De Anza

STEFANI THOMAS, PhD, DABCC, NRCC
UNIVERSITY OF MINNESOTA

The mutational status of a solid tumor can predict the therapeutic efficacy of a specific drug in a molecularly defined subset of patients. Targeted therapies are available to treat advanced (stage II – IV) ovarian cancer with mutations in BRCA1/2 genes. Unfortunately, there is considerable inter-patient heterogeneity in BRCA1/2–based determinations of drug treatment sensitivity. Determining the proteome-level mechanisms of drug treatment sensitivity could enhance our ability to select the ovarian cancer patient populations that would benefit the most from these targeted therapies, consequently improving survival and overall treatment response. Our laboratory is applying mass spectrometry-based proteomics to identify protein signatures of drug treatment sensitivity and subsequent patient stratification for treatment. This presentation will provide an overview of the experimental models and analytical approaches that we are utilizing toward a long-term goal of identifying prognostic protein biomarkers of drug treatment sensitivity in patients with high-grade serous ovarian cancer.

Moderated by: LAURA SANCHEZ, PhD

N-linked Glycans in Human Disease: From New Tools to Translational and Preclinical Studies

Wednesday 9:45-10:15 @ De Anza

PEGGI ANGEL, PhD
MUSC PROTEOMICS CENTER

N-glycosylation plays a significant role in immune cell recruitment, influences disease progression and outcome and response to therapy. Here, we discuss simplified workflows capable of reporting N-glycan expression patterns in tissues, cells and biofluids. We present translational and pre-clinical work investigating glycosylation patterns in cardiovascular disease, cancer risk and cancer. A long-term goal is to leverage glycosylation patterns to non-invasively monitor disease status and therapeutic efficacy.

Moderated by: LAURA SANCHEZ, PhD

Proteins in Space: Statistical Approaches to Understand the Spatial Organization and Structure of Proteins in Complex Tissues

Thursday 8:00-8:55 @ De Anza

BARBARA ENGELHARDT, PHD

PRINCETON UNIVERSITY

Spatial single-cell technologies have enabled the comprehensive study of complex tissues and organs at a single-cell level. However, the statistical methods to perform analyses have not kept pace with the technologies for measuring spatially-resolved genes and proteins in tissue samples. In this talk, I will present two methods that allow the study of spatially-resolved proteomics data. First, nonnegative spatial factorization develops a low-dimensional representation of the spatially-resolved samples to identify variation in protein expression across space. Second, Gaussian process spatial alignment allows multiple slices from different technologies to be aligned to a known or unknown common coordinate system, enabling the construction of a tissue atlas from disparate samples of that tissue type.

Moderated by:

SHANNON HAYMOND, PHD
NORTHWESTERN UNIVERSITY FEINBERG SCHOOL OF MEDICINE

Statistical Considerations for Biomarker Discovery Experiments: From a Model Organism to Clinical Study

Thursday 9:15-10:00 @ De Anza

MEENA CHOI, PHD

GENENTECH

Quantitative mass spectrometry-based proteomics is a technology of growing importance in biological and clinical research. As modern quantitative mass spectrometry-based proteomics workflows become complex and diverse, it requires statistical considerations, methods, and computational tools for experimental planning and data analysis to reduce bias and inefficiencies and maximize the reproducibility of results. This talk will highlight statistics essential for proteomics experiments and considerations for designing a proteomics experiment for biological research biomarker discovery and issues related to large quantitative proteomic datasets generated by diverse types of acquisition of proteomics. Also, this talk will present the comprehensive mass-spectrometry proteomics-based strategy and case studies for the clinical protein biomarker development, from a model organism to large-scale clinical samples, with rigorous experimental design and statistical analysis.

Moderated by: **SHANNON HAYMOND, PHD**

Adapting to the New Normal: Navigating the Rollercoaster of SARS-CoV-2 Testing Needs While Building Long Term Capabilities

Thursday 10:15-11:00 @ De Anza

PATRICK MATHIAS, MD, PHD

UNIVERSITY OF WASHINGTON

The COVID-19 pandemic has been a challenging time for clinical laboratories, but it has also improved awareness of the important role labs play across health care and public health domains. The University of Washington Department of Laboratory Medicine and Pathology has worked to establish diagnostic SARS-CoV-2 testing as a widely available resource over a large footprint of our state by combining expertise in laboratory testing and informatics with community partnerships to expand testing access and convenience. Over the course of the pandemic the needs of our laboratories as well as of the public and of public health authorities have evolved. Staffing and supply chain challenges have placed a greater emphasis on efficiency and productivity. The need for genomic surveillance of the evolving virus has required a close integration of clinical workflows with research-focused genomics workflows. Increased demand for services coupled with record high positivity rates during the omicron surge has challenged our model for expanding capacity using sample pooling. This presentation will describe the evolution of our testing program and infrastructure amidst these challenges, highlighting the informatics capabilities we needed to adapt. Finally, we will discuss what needs and capabilities are likely to persist beyond the pandemic and how the lessons we have learned can better equip laboratories to improve the health of the public.

Moderated by: **SHANNON HAYMOND, PHD**

Addressing Hurdles in Clinical Translation of Targeted Proteomics

Friday 9:00-10:00 @ De Anza

JEFFREY WHITEAKER, PHD

FRED HUTCHINSON CANCER RESEARCH CENTER

Quantifying proteins and post-translational modifications will improve precision medicine, but several hurdles remain to adopting proteomics to the clinical laboratory. Dr. Whiteaker will discuss successes and remaining challenges for incorporating targeted proteomic measurements in clinical trials and other clinical applications.

Moderated by:

STEFANI THOMAS, PHD, DABCC, NRCC
UNIVERSITY OF MINNESOTA

TIMOTHY COLLIER, PHD
QUEST DIAGNOSTICS

Newborn Screening by Mass Spec Meets Newborn Screening by DNA Sequencing

Friday 10:00-10:30 @ De Anza

Michael Gelb, PhD
University of Washington

Our laboratory has been developing tandem mass spectrometry (MS/MS) for worldwide expansion of newborn screening (NBS) panels to include an ever-increasing collection of treatable genetic diseases. There is widespread discussion on the use of whole genome and whole exome DNA sequencing in population-wide NBS. The intersection of biochemical- and DNA-based NBS is an interesting topic now under heavy discussion.

We will highlight the development of liquid chromatography-MS/MS (LC-MS/MS) for multiplex NBS of a large panel of treatable genetic diseases in newborns. Next generation sequencing (NGS) is also employed currently as a second-tier analysis after LC-MS/MS assays. We will also illustrate how it is possible to carry out first-tier NGS followed by second-tier LC-MS/MS NBS.

LC-MS/MS is used together with enzyme substrates and biomarkers to monitor the activity of a large collection of enzymes and to measure the abundance of biomarkers in dried blood spots on NBS cards. We will focus on multiplex methods and then zoom in one a more detailed analysis of one disease called metachromatic leukodystrophy (MLD). We carried out a pilot MLD NBS study and determined that the rate of false positives out of 28,000 newborns screened is essentially zero showing the power of LC-MS/MS for NBS of this lysosomal storage disorder. In the second arm of the study, we have been measuring the activity of the enzyme relevant to MLD on a large collection of gene variants that are found in allele databases and for which no pathogenic information is reported. We show how we can integrate these efforts to provide for a highly efficient NBS program for MLD.

We screened ~28,000 newborns for elevated sulfatide lipid, the biomarker that is relevant to MLD and found 180 high sulfatide newborns. These were submitted to an assay of the activity of the relevant enzyme, arylsulfatase A, and all but two showed normal levels of activity. DNA sequencing was carried out on 2 newborns, one with 0% and one with 8% of normal ARSA activity. The newborn with 0% activity was confirmed to have MLD, the other was shown to not have MLD. On the DNA front, we created a phenotype matrix that allows one to input the ARSA enzymatic activity of each variant to provide a composite genotype, and to make a prediction of the phenotype associated with this genotype. We show that this method is 83% accurate at predicting the true set of phenotypes observed in MLD patients.

Massively multiplexed NBS of genetic diseases in newborns is possible using LC-MS/MS, and when used with second-tier NGS leads to a successful NBS platform. We show that it is also possible to carry out NGS as a first-tier NBS step and to clarify the results with second-tier biochemical assays based on LC-MS/MS. Thus LC-MS/MS meets DNA and DNA meets LC-MS/MS, and this provides a framework for the future employment of both LC-MS/MS and NGS in expansion of population based NBS.

Moderated by: **Stefani Thomas, PhD, DABCC, NRCC and Tim Collier, PhD**

Utilization of Mass Spectrometry to Discover and Develop Novel Biomarkers to Support Drug Development

Friday 10:50-11:20 @ De Anza

VERONICA ANANIA, PHD

GENENTECH

Biomarkers play an important role in the drug development process including providing necessary insights into target engagement, dose selection and mechanism of action of candidate therapeutics. LC-MS is uniquely positioned to enable accurate quantitation of both small and large molecule biomarker candidates, however, the process of going from biomarker discovery to a multiplexed targeted MRM panel in clinical samples is long and resource intensive. Moreover, biomarker candidates often fail to replicate when tested in large clinical cohorts. Recent advances in data-independent MS (DIA-MS) have made this technology more accessible and certain benefits of DIA-MS including reproducible label-free analysis of hundreds of samples, ability to capture low abundance ions over a high dynamic range, and deep proteome coverage makes this technology well suited to streamline translational proteomics. One major hurdle for using DIA-MS to support drug development is that the quantitative range for most DIA-MS methods has not been well characterized and thus, quantitative conclusions drawn by prior studies that have employed this approach have been controversial. Here, we describe challenges associated with applying DIA-MS methods to address questions associated with clinical development and introduce best practices for establishing quantitative criteria for DIA-MS approaches in clinical trial samples. Results and lessons learned from both discovery and targeted clinical biomarker studies will be discussed and a model for a more streamlined biomarker development workflow that conserves resources and provides more comprehensive proteomic information from clinical trial samples will be discussed.

Moderated by: **Stefani Thomas, PHD, DABCC, NRCC and Tim Collier, PhD**

Round Table Discussion Groups

Wednesday 7:00 -7:45am @ De Anza Foyer

(1) On Being a Reviewer
with Tim Garrett

(2) Follow-on Discussion on *How to Convince Admin That THEY Need to Buy a Mass Spectrometer*
with Joshua Hayden

(3) Shortcomings / Limitations of Antinuclear Antibodies (ANA) : Stark Disparities – Multiplex vs. Immunofluorescence Assay
with Mahesheema Ali

Thursday 7:00 -7:45am @ De Anza Foyer

(1) JMSACL co-Editors-in-Chief : Current and Interested Editors and Reviewers Round-Up
with Kara Lynch &Tim Garrett

(2) Meet a Clinical Chemist
with Shannon Haymond and Stephen Master

(3) Tissue Imaging Satellite Conference Discussion
with Michael Angelo & David Herold

Industry Workshops

Pre-conference Registration Required for All Industry Workshops
includes box lunch

Wednesday 12:00 – 13:00

Thermo Fisher Scientific
@ De Anza 1 (Track 1)

Mass Spectrometry-based Tools Empowering the State of Healthcare – Today and Tomorrow

Part 1 : From the OR to the Microbiology Lab: Advanced Development of the MasSpec Pen for Broad Clinical Use

Mary King, PhD Candidate – Prof. Livia S. Eberlin Group
The University of Texas at Austin

Part 2 : Enhanced LC-MS Detection of Severe Acute Respiratory Syndrome Coronavirus 2 (SARS-CoV-2) Peptides

Richard Gibson, PhD
Thermo Fisher Scientific

Thermo Fisher Scientific
@ De Anza 2 (Track 2)

A Wide Range of Mass Spec Solutions for the Clinical Laboratory : From Flexibility for LDTs to a New Fully Automated Clinical Analyser for Easy Implementation

Part 1 : LC-MS/MS Adoption in the Clinical Laboratory : Current Trends and Future Perspectives

Kara Lynch, PhD
University of California San Francisco

Part 2 : Measuring Vitamin D2 and D3 on the Cascadion™ Fully Automated LC-MS/MS System : From Installation to Result Reporting

Philip Sobolesky, PhD
Santa Clara Valley Medical Center

Agilent Technologies
@ De Anza 3 (Track 3)

Keeping an "Eye on" Mobility Technologies for Clinical Research Laboratory Workflows

Frederick Strathmann, PhD, MBA
MOBILion Systems

Automated Rapid Throughput Quantitative Proteomics of Amniotic Fluid Utilizing Agilent AssayMap Bravo and UHPLC-MS/MS

Anita Vinjamuri, Bachelors of Science in Chemistry
University of California Davis

Armin Oloumi, BS in Biochemistry
University of California, Davis

Industry Workshops, continued

Thursday 12:00 – 13:00

Thermo Fisher Scientific
@ De Anza 1 (Track 1)

Advances in High Throughput Mass Spectrometry – Novel Front-End Solutions Driving Routine Clinical TDM Analysis

Part 1 : A Therapeutic Drug Monitoring Assay of Immunosuppressants using TurboFlow LC and High-Resolution Mass Spectrometry

Y. Ruben Luo, PhD, DABCC
Stanford University

Part 2 : High Throughput Paper-Spray Mass Spectrometry (PS-MS) for Validated Clozapine Therapeutic Drug Monitoring in Human Serum

Chris Gill, Ph.D.
Vancouver Island University

Indigo BioAutomation
@ De Anza 2 (Track 2)

Collabalytics – Using Data Analytics to Elevate Performance and Foster Collaboration in the Laboratory

Jim Edwards
Indigo BioAutomation

Shimadzu
@ De Anza 3 (Track 3)

Do More with Less: Ease the Burden of Staff Shortages while Simultaneously Boosting Lab Productivity with Ultra-Fast Mass Spectrometry

Scott Kuzdzal, Ph.D., Analytical Chemistry
Shimadzu Scientific Instruments

Scaling Up while Scaling Down: New Era for Plasma/Serum Proteomics

Benoit Fatou, PhD
Boston Children's Hospital, Harvard Medical School

Podium Presentations

Scientific Session 1 : Wednesday

①	②	③	④
Track 1 in **De Anza 1 Endogenous Small Molecules**	Track 2 in **De Anza 2 Glycomics and proteomics of SARS-CoV2 Immunology**	Track 3 in **De Anza 3 Taking it to the Top: Reference Methods and Materials**	Track 4 in **Bonzai CLOSED**
Chair: B. Keevil 2nd: J. Pagaduan	Chair: R. Bearden	Chair: K. Phinney 2nd: A. Nelson	

Session 1 on Wednesday @ 14:00 (a)

Track ①: 1.1.a on Wed @ 14:00

⊙ **Ultrafast Integrated UPLC-MS/MS Second-tier Newborn Screening for Inborn Errors of Propionate, Cobalamin, and Methionine Metabolism**

Joshua Dubland
The University of British Columbia & BC Children's Hospital
Young Investigator Grantee

Joshua Dubland

Track ②: 1.2.a on Wed @ 14:00

⊙ **Rational Design of Serological Diagnostics Using Immunoprecipitation-Targeted Proteomic Assays**

Andrei Drabovich
University of Alberta
Young Investigator Grantee

Andrei Drabovich

Track ③: 1.3.a on Wed @ 14:00

⊙ **Internal Standard Approaches: Stable Isotope-labeled Intact, Winged or Peptide**

Kayla Moehnke
Mayo Clinic

Kayla Moehnke

Session 1 on Wednesday @ 14:20 (b)

Track ①: 1.1.b on Wed @ 14:20

◎ **Newborn Screening for Vitamin B12 Deficiency: Is it Justified? Is it Possible?**

Bojana Rakic
BC Children's Hospital, Vancouver, Canada

Bojana Rakic

Track ②: 1.2.b on Wed @ 14:20

◎ **Afucosylated IgG Responses to BNT162b2 mRNA Vaccine Against Sars-CoV-2 Differ in Naïve and Antigen-experienced Individuals**

Tamas Pongracz
Leiden University Medical Center
Young Investigator Grantee

Tamas Pongracz

Track ③: 1.3.b on Wed @ 14:20

◎ **Development of a Candidate Reference Measurement Procedure for Urine Albumin Using LC-MS/MS**

Jesse Seegmiller
University of Minnesota

Jesse Seegmiller

Session 1 on Wednesday @ 14:40 (c)

Track ①: 1.1.c on Wed @ 14:40

◎ **Urine Free Cortisol by LC-MS/MS: Monitoring Known Interferences as Potential Culprits of Iatrogenic Cushing's Syndrome**

Julie Ray
ARUP Labs

Julie
Ray

Track ②: 1.2.c on Wed @ 14:40

◎ **A Targeted Multiplex MRM LC-MS/MS Immunoassay for Characterisation of Antibody Response to SARS CoV2 Vaccination**

Ivan Doykov
University College London

Ivan
Doykov

Track ③: 1.3.c on Wed @ 14:40

◎ **Get Your Reference Right!: Synthesis, Certification, and Confirmation of the First Reference Material for (-)-trans-11-nor-9-Carboxy-Δ9-THC beta-d-glucuronide**

Raymond Suhandynata
University of California, San Diego

Raymond
Suhandynata

① Track 1 in **De Anza 1**
Rapid Detection and Quantitation via Ambient Ionization
Chair: V. Samara

② Track 2 in **De Anza 2**
Tissue imaging – Dynamics and Mapping
Chair: P. Angel
2nd: J. Weatherill

③ Track 3 in **De Anza 3**
Starting Right – The Life of a Sample from Tube to Bench
Chair: M. Chen

④ Track 4 in **Bonzai**
Practical Training
Chair: Self-Chaired

Session 2 on Wednesday @ 15:15 (a)

Track ①: 2.1.a on Wed @ 15:15

◎ **Rapid Detection of Anticoagulants Using Coated Blade Spray Mass Spectrometry**

Briana Fitch
University of California San Francisco
Young Investigator Grantee

Briana Fitch

Track ②: 2.2.a on Wed @ 15:15

◎ **Mass Spectrometry Imaging of Glutathione Biosynthesis Heterogeneity Using Multiple Infusion Start Times and IR-MALDESI**

Allyson Mellinger
North Carolina State University

Allyson Mellinger

Track ③: 2.3.a on Wed @ 15:15

◎ **Tracking the Stability of Clinical Blood Plasma Proteins with ΔS-Cys-Albumin—a Dilute-and-shoot LC-MS-based Marker of Specimen Exposure to Thawed Conditions**

Chad Borges
Arizona State University

Chad Borges

Track ④: 2.4.a on Wed @ 15:15

This is a 1-hour TRAINING session.

◎ **Oral Fluid as Alternative Matrix for Drug Testing: Clinical Utility and Method Development**

Adina Badea
Lifespan Health/Rhode Island Hospital & The Warren Alpert Medical School of Brown University

Adina Badea

Session 2 on Wednesday @ 15:35 (b)

Track ①: 2.1.b on Wed @ 15:35

◎ **Concomitant Quantification of Methotrexate and its Metabolites in Serum Samples via Fully Automated Coated Blade Spray-Tandem Mass Spectrometry Workflow**

 **German Gomez**

Restek

German Gomez

Track ②: 2.2.b on Wed @ 15:35

◎ **Mass Spectrometry Histochemistry of Human FFPE Material Getting Ready for the Clinic**

 **Peter Verhaert**

ProteoFormiX

 Peter Verhaert

Track ③: 2.3.b on Wed @ 15:35

◎ **The Decisive Role of Pre-analytical Sample Handling When Investigating the Lipidome in Clinical Research**

 **Daniel Kratz**

Institute of Clinical Pharmacology, Pharmazentrum Frankfurt/Zafes, University Hospital of Goethe-University

Young Investigator Grantee

Daniel Kratz

Session 2 on Wednesday @ 15:55 (c)

Track ①: 2.1.c on Wed @ 15:55

◎ **Rapid Urine Drug Testing by Direct Analysis in Real-Time (DART)-Mass Spectrometry**

Ibrahim Choucair
Yale School of Medicine

Ibrahim Choucair

Track ②: 2.2.c on Wed @ 15:55

◎ **Improving Accuracy and Speed of MALDI MSI for Clinical Translation**

Sankha (Bobby) Basu
Brigham and Women's Hospital

Sankha (Bobby) Basu

Track ③: 2.3.c on Wed @ 15:55

◎ **The Essence of a Streamlined Process for Large-Scale Quantitative Protein Mass Spectrometry**

Renee Ruhaak
LUMC

Renee Ruhaak

Scientific Session 3 : Wednesday

①
Track 1 in **De Anza 1**
Steroids – Methods and Matricies
Chair: C. Knezevic

②
Track 2 in **De Anza 2**
Proteomics – New Methods and Continuous Improvement
Chair: M. DeMarco

③
Track 3 in **De Anza 3**
Data – Integrating, Mapping and Reporting
Chair: P. Vanderboom

④
Track 4 in **Bonzai**
Practical Training
Chair: Self-Chaired

Session 3 on Wednesday @ 16:30 (a)

Track ①: 3.1.a on Wed @ 16:30

◎ **Application of a Novel Steroidomic LC-MS/MS Method for Investigating Metabolic Alterations in Patients with Non-Alcoholic Fatty Liver Disease**

Federico Ponzetto
University of Turin

Federico Ponzetto

Track ②: 3.2.a on Wed @ 16:30

◎ **Proteomic Analyses of Malaria Drug Resistance**

Annie Moradian
Precision Biomarker Laboratories

Annie Moradian

Track ③: 3.3.a on Wed @ 16:30

◎ **Data Integration Pipeline for Multi-Batch Untargeted Lipidomics Studies**

Khyati Mehta
Cincinnati Children's Hospital/University of Cincinnati
Young Investigator Grantee

Khyati Mehta

Track ④: 3.4.a on Wed @ 16:30

This is a 1-hour TRAINING session.

◎ **Transition Ratio Monitoring for the Masses**

Russell Grant
Labcorp

Russell Grant

Session 3 on Wednesday @ 16:50 (b)

Track ①: 3.1.b on Wed @ 16:50

◎ **Ion Mobility-Mass Spectrometry: Targeted Steroid Applications in the Clinical Lab**

 **Christopher Chouinard**
Florida Institute of Technology

Christopher Chouinard

Track ②: 3.2.b on Wed @ 16:50

◎ **Integrating Lipidomics and Proteomics for Increased Diagnostic Accuracy in Prostate Cancer**

 **Lee Gethings**
Waters

 Lee Gethings

Track ③: 3.3.b on Wed @ 16:50

◎ **The Chemistry of Our Blood: Mapping Circulating Molecules to the Landscape of Human Health and Disease**

 **Nicholas Bevins**
Sapient Bioanalytics

Nicholas Bevins

Track ①: 3.1.c on Wed @ 17:10

◎ **Salivary Cortisone in the Assessment of the HPA Axis**

Brian Keevil
University Hospital of South Manchester

Brian Keevil

Track ②: 3.2.c on Wed @ 17:10

◎ **High Sensitivity Measurement of Thyroglobulin Using Conventional Flowrate LC-MS/MS**

Mark Kushnir
ARUP Institute for Clinical & Experimental Pathology

Mark Kushnir

Track ③: 3.3.c on Wed @ 17:10

◎ **Using Data Independent Acquisition to Inform the Development of Cerebrospinal Fluid Triple Quadrupole Assays**

Deanna Plubell
University of Washington

Deanna Plubell

①
Track 1 in **De Anza 1**
Microbiology -
Identification and
Metabolomics
Chair: J. Whitman

②
Track 2 in **De Anza 2**
Monitoring mAb
Chair: Y. Luo
2nd: R. Kumar

③
Track 3 in **De Anza 3**
Making the Most of
the Data We Harvest
Chair: P. Ladwig
2nd: J. Kemp

④
Track 4 in **Bonzai**
Practical Training
Chair: Self-Chaired

Session 4 on Thursday @ 14:00 (a)

Track ①: 4.1.a on Thu @ 14:00

◎ **Identification of the Subspecies from Mycobacterium Abscessus Complex by MALDI-TOF MS and Machine Learning Approach**

David Rodriguez-Temporal
Hospital General Universitario
Gregorio Marañón, Madrid, Spain
Young Investigator Grantee

David
Rodriguez-Temporal

Track ②: 4.2.a on Thu @ 14:00

◎ **Therapeutic Antibody Monitoring: the Case of Infliximab**

Maria Willrich
Mayo Clinic

Maria
Willrich

Track ③: 4.3.a on Thu @ 14:00

◎ **Development and Implementation of a Web-based Application for Urine Drug Testing and Improving Resulting Workflow**

Abed Pablo
University of Washington

Abed
Pablo

Track ④: 4.4.a on Thu @ 14:00

This is a 1-hour TRAINING session.
◎ **Working Smarter Not Harder on Sample Prep**

Matthew Crawford
Labcorp

Matthew
Crawford

Session 4 on Thursday @ 14:20 (b)

Track ①: 4.1.b on Thu @ 14:20

◎ **Unique Chemical Biology Tools for Metabolomics Analysis – Exploring Gut Microbiota Metabolism**

Daniel Globisch
Uppsala University

Daniel Globisch

Track ②: 4.2.b on Thu @ 14:20

◎ **Development and Validation of a Bioanalytical Method for Certolizumab Pegol (Cimzia) Drug Using Surface Plasmon Resonance**

Brett Holmquist
Labcorp

Brett Holmquist

Track ③: 4.3.b on Thu @ 14:20

◎ **Machine Learning-based Fragment Selection Improves Performance of Qualitative PRM Assays for Viral Pathogen Screening**

Patrick Vanderboom
Mayo Clinic

Patrick Vanderboom

Session 4 on Thursday @ 14:40 (c)

Track ①: 4.1.c on Thu @ 14:40

◎ **Investigating Microbial Cooperation and Metabolic Communication During Clostrioles Difficile Infection Using Imaging Mass Spectrometry**

Boone Prentice
University of Florida

Boone
Prentice

Track ②: 4.2.c on Thu @ 14:40

◎ **Measurement of Therapeutic and Anti-drug Antibodies Using Biolayer Interferometry**

Kara Lynch
University of California San Francisco

Kara
Lynch

Track ③: 4.3.c on Thu @ 14:40

◎ **Isotopic Distribution Calibration for Mass Spectrometry**

Anthony Maus
Mayo Clinic

Anthony
Maus

① **Track 1 in De Anza 1**
Vitamin D
Chair: J. Ray
2nd: A. Bajaj

② **Track 2 in De Anza 2**
Drugs of Abuse
Chair: M. Budelier
2nd: I. Metushi

③ **Track 3 in De Anza 3**
Late-Breaking
Submissions A
Chair: K. Johnson-Davis

④ **Track 4 in Bonzai**
Practical Training
Chair: Self-Chaired

Session 5 on Thursday @ 15:15 (a)

Track ①: 5.1.a on Thu @ 15:15

◉ **Learning from Vitamin D Binding Protein**

Andy Hoofnagle
University of Washington

Andy Hoofnagle

Track ②: 5.2.a on Thu @ 15:15

◉ **Analysis of Suspected Opioid Overdose Samples Using High Resolution Mass Spectrometry (HRMS) for Public Health Opioids Biosurveillance Program in South Carolina**

James LaPalme
SC PHL

James LaPalme

Track ③: 5.3.a on Thu @ 15:15

◉ **Detection of Mycobacterial from Species to Sub-strain Level with Pathogen-derived Peptidomes**

Tony Hu
Tulane University School of Medicine

Tony Hu

Track ④: 5.4.a on Thu @ 15:15

This is a 1-hour TRAINING session.

◉ **Introduction to Clinical Proteomics Using Mass Spectrometry**

Timothy Collier
Quest Diagnostics

Timothy Collier

Session 5 on Thursday @ 15:35 (b)

Track ①: 5.1.b on Thu @ 15:35

◎ **Free 25-hydroxy Vitamin D Measured by LC-MS/MS in Healthy Adults and Individuals with Kidney Disease**

Mark Kushnir
ARUP Institute for Clinical & Experimental Pathology

Mark Kushnir

Track ②: 5.2.b on Thu @ 15:35

◎ **Opioid Overdose Harm Reduction Drug Checking and Illicit Drug Supply Surveillance by On-site Paper Spray Mass Spectrometry (PS-MS)**

Chris Gill
Vancouver Island University

Chris Gill

Track ③: 5.3.b on Thu @ 15:35

◉ **Novel First Trimester Biomarkers of Fetal Growth Restriction Identified by Next Generation Proteomics of Maternal Plasma**

Nandor Gabor Than
Research Centre for Natural Sciences, Budapest, Hungary

Nandor Gabor Than

Session 5 on Thursday @ 15:55 (c)

Track ①: 5.1.c on Thu @ 15:55

◎ **Clinical Utility of Measuring the Vitamin D Metabolome including 24,25-(OH)2D3 by LC-MS/MS**

Martin Kaufmann
Queen's University

Martin Kaufmann

Track ②: 5.2.c on Thu @ 15:55

◎ **Opiate Drug Testing Does Not have to be a "pain in the MLS": How a Second-generation LC-MS/MS Assay Improved Workflow for Drug Testing and Sign-out**

Hsuan-Chieh (Joyce) Liao
University of Washington

Hsuan-Chieh (Joyce) Liao

Track ③: 5.3.c on Thu @ 15:55

◎ **Precise Quantitation of PTEN by Immuno-MRM: A Tool to Resolve the Breast Cancer Biomarker Controversy**

Christoph Borchers
Jewish General Hospital, McGill University Montreal, QC, Canada

Christoph Borchers

① Track 1 in **De Anza 1**
Multiplexing Small Molecules and Glycans
Chair: A. Mellinger

② Track 2 in **De Anza 2**
Top-Down Protein Characterization
Chair: T. Collier

③ Track 3 in **De Anza 3**
Late-Breaking Submissions B
Chair: G. McMillin
2nd: D. Liu

④ Track 4 in **Bonzai**
Practical Training
Chair: Self-Chaired

Session 6 on Thursday @ 16:30 (a)

Track ①: 6.1.a on Thu @ 16:30

◎ **Targeting Cellular Nucleotide Biosynthesis Using Ultrahigh-Performance Liquid Chromatography Coupled with High Resolution Mass Spectrometry**

Xueheng Zhao
Cincinnati Children's Hospital Medical Center

Xueheng Zhao

Track ②: 6.2.a on Thu @ 16:30

◎ **Discovery of a Biomarker for beta-Thalassemia by Fourier Transform Ion Cyclotron Resonance Mass Spectrometry and Proton Transfer Reaction- Parallel Ion Parking**

Yuan Lin
Florida State University
Young Investigator Grantee

Yuan Lin

Track ③: 6.3.a on Thu @ 16:30

◎ **Mitochondrial and Metabolic Dysfunctions in Neurons Carrying the 22q11.2 Deletion**

Wojciech Michno
Stanford University / University College London

Wojciech Michno

Track ④: 6.4.a on Thu @ 16:30

This is a 1-hour TRAINING session.

◎ **Free Climb or Free Solo: Guidance and Accounts of Establishing Reference Intervals for LDT Mass Spectrometry Assays**

Elizabeth Frank & Kelly Doyle
University of Utah Health / ARUP Laboratories

Elizabeth Frank

Session 6 on Thursday @ 16:50 (b)

Track ①: 6.1.b on Thu @ 16:50

○ **Tumor Associated Carbohydrate Antigens (TACAs) as Promising Targets for the Development of Immunotherapy for Colorectal Cancer**

Katarina Madunić
Leiden University Medical Centre
Young Investigator Grantee

Katarina
Madunić

Track ②: 6.2.b on Thu @ 16:50

◎ **Capillary Zone Electrophoresis Coupling with High-resolution Mass Spectrometry for Top-down Structural Characterization of Hemoglobin Variants**

Y. Ruben Luo
Stanford University

Y. Ruben
Luo

Track ③: 6.3.b on Thu @ 16:50

◎ **Correlation of Genotyping and Phenotyping with the IGF-1 assays for Growth Hormone disorders**

Ravinder Singh
Mayo Clinic

Ravinder
Singh

Session 6 on Thursday @ 17:10 (c)

Track ①: 6.1.c on Thu @ 17:10

◎ **Serum Bile Acid Profiling by UHPLC-MS/MS Predicts Cholesteric Pruritus Reduction in Maralixibat Treated Patients with Bile Salt Export Pump Deficiency**

Kenneth Setchell
Cincinnati Children's Hospital Medical Center

Kenneth Setchell

Track ②: 6.2.c on Thu @ 17:10

◎ **Detection of an Unusual IGF-2 Species Using High-resolution Mass Spectrometry During a Routine IGF-1 Clinical Assay**

Ievgen Motorykin
Quest Diagnostics

Ievgen Motorykin

Track ③: 6.3.c on Thu @ 17:10

◎ **Mapping the 3 D Proteome Using Surface Accessibility in Animal Models of Disease**

John Yates
Scripps Research Institute

John Yates

Poster Presentations

Assays Leveraging MS

Poster #11a : *Wednesday : 11:00* — Assays Leveraging MS

◎ **Serum-Based Free Light Chain Posttranslational Modification and Multimer Characterization Assay**
Ira Miller - *Mayo Clinic*

Ira Miller

Poster #23a : *Wednesday : 11:00* — Assays Leveraging MS

◎ **Determination of Free Testosterone from Human Serum Using BioSPME Sample Preparation Prior to LC-MS/MS Analysis**
M. James Ross - *MilliporeSigma*

M. James Ross

Poster #25a : *Wednesday : 11:00* — Assays Leveraging MS

◎ **An Automated Procedure for Urine Catecholamines that Increases Throughput by Multiplexed LC-MS/MS**
John Ross - *Quest Diagnostics*

John Ross

Poster #37a : *Wednesday : 11:00* — Assays Leveraging MS

◎ **Assessing Cardiovascular Risk Using LC-MS/MS for Simultaneous Determination of 4 Ceramides and 3 Phosphatidylcholines in Serum**
Haiqing Ding - *Quest Diagnostics / Cleveland Heart Lab*

Haiqing Ding

Poster #39b : *Wednesday : 13:00* — Assays Leveraging MS

◎ **A Peptide Enrichment and Mass Spectrometry-based Workflow for the Absolute Quantitation of SARS-CoV-2**
Richard Gibson - *Thermo Fisher Scientific*

Richard Gibson

Poster #53b : *Wednesday : 13:00* — Assays Leveraging MS

◎ **Better Together: A Novel Ion-pairing Method for an Extended Pathway, Low-Volume, Plasma Catecholamine Assay by LC-MS/MS**
Andrzej Szczesniewski - *Agilent Technologies*

Andrzej Szczesniewski

Poster #2a : *Thursday : 11:00*   Assays Leveraging MS

◎ **Quantitative Analysis of Vitamin B1 in Whole Blood on a Multiplex-ESI-LC-MS/MS Platform**
Preejith Vachali - *ARUP*

Preejith
Vachali

Poster #36a : *Thursday : 11:00*   Assays Leveraging MS

◎ **Middle-Up Approach for Therapeutic Monoclonal Antibodies (t-mAbs) Monitoring in Human Serum Using LC-HRAM-MS**
Yvonne Song - *Thermo Fisher Scientific*

Yvonne
Song

Poster #12b : *Thursday : 13:00*   Assays Leveraging MS

◎ **Analysis of Estrogens in Plasma with Rapid Chromatography and Reduced Sample Volume**
Xiaohong Chen - *SCIEX*

Xiaohong
Chen

Poster #16b : *Thursday : 13:00*   Assays Leveraging MS

◎ **Utilizing LC-MS/MS for the Clinical Quantitation of Testosterone Using Dried Blood Spot and How They Compare to Serum**
Zane Hauck - *ZRT Laboratory*

Zane
Hauck

Poster #18b : *Thursday : 13:00*   Assays Leveraging MS

◎ **Small Footprint, Big Screen: Developing a Broad-spectrum Drug Screen for Over 100 Analytes in Urine/Serum Utilizing the Agilent Ultivo LC-MS/MS**
Kathryn Smith - *ARUP*

Kathryn
Smith

Poster #22b : *Thursday : 13:00*   Assays Leveraging MS

◎ **Detection of Metabolic Urine Organic Acids via Atmospheric Pressure Chemical Ionization Gas Chromatography Mass Spectroscopy in Multi-Reaction Monitoring Mode**
Devanjith Ganepola - *McMaster University*

Devanjith
Ganepola

Poster #24b : *Thursday : 13:00*   Assays Leveraging MS

◎ **Analytical Validation of a Glycoproteomics-based Assay for Determining the Likelihood of Malignancy Among Newly Diagnosed Pelvic Tumors**
Maurice Wong - *InterVenn Biosciences*

Maurice
Wong

Poster #34b : *Thursday : 13:00*   Assays Leveraging MS

◎ **A Story of Discovery and Translation: Advances in Plasma Aβ Biomarkers for Alzheimer's Disease**
Scott Kuzdzal - *Shimadzu Scientific Instruments*

Scott Kuzdzal

Poster #40b : *Thursday : 13:00*   Assays Leveraging MS

◎ **Simultaneous Determination of Tryptophan and its Kynurenine Pathway Metabolites by LC-MS/MS in Serum**
Mehmet Balcı - *Sem Laboratuar Cihazları Pazarlama San. Ve Tic. A.Ş*

Mehmet Balcı

Poster #46b : *Thursday : 13:00*   Assays Leveraging MS

◎ **Development and Validation of a Sensitive LC-MS/MS Method for the Determination of Atovaquone in Human Plasma Samples**
Lily Olayinka - *Baylor College of Medicine*
Young Investigator Grantee

Lily Olayinka

Cases in Clinical MS

Poster #39a : *Wednesday : 11:00*   Cases in Clinical MS

◎ **Health Surveillance Panel Multiplexed MRM-based Plasma Protein Assay for the Identification of Multiple Biomarkers of Disease Severity in Human Coronary Disease**
Esthelle Hoedt - *Cedars Sinai/ Precision Biomarker Labs*

Esthelle Hoedt

Poster #47a : *Wednesday : 11:00*   Cases in Clinical MS

◎ **A Novel LC-MS/MS Method for Direct Analysis of Underivatized Amino Acid in Human Plasma**
Amol Kafle - *Restek*

Amol Kafle

Poster #1b : *Wednesday : 13:00*   Cases in Clinical MS

◎ **Measurement of Hepcidin-25 Using Commercial LC-MS/MS Assay to Predict Iron Absorption in Inflammatory Bowel Disease**
Fabian Simon - *Immundiagnostik AG, Bensheim, Germany*

Fabian Simon

Data Analytics

Poster #37b : *Wednesday : 13:00*   Data Analytics

◎ **Ion Ratio Evaluation Using a Linear Regression-based Approach**
Christopher Koch - *Sanford Health*
Young Investigator Grantee

Christopher
Koch

Emerging Technologies

Poster #43b : *Wednesday : 13:00*   Emerging Technologies

◎ **Automating Low-Volume 96-Well SPE Assays for Forensic and Clinical Toxicology Prior to UHPLC-MS/MS Analysis**
Adam Senior - *Biotage GB ltd*

Adam
Senior

Poster #40a : *Thursday : 11:00*   Emerging Technologies

◎ **Direct Quantitation of Phosphatidylethanol (Peth) in Volume-Controlled Dried Blood Spots Using the Fully Automated Transcend DSX-1 System**
Jingshu Guo - *Thermo Fisher Scientific*

Jingshu
Guo

Poster #42a : *Thursday : 11:00*   Emerging Technologies

◎ **Shaping the Future of Estrogen Analysis: Assessing Separation and Detection of Metabolites by Ion Mobility-Mass Spectrometry**
James Weatherill - *The University of Edinburgh*
Young Investigator Grantee

James
Weatherill

Glycomics

Poster #13b : *Wednesday : 13:00*   Glycomics

◎ **New NIST Reference Material Helps Biomanufacturers Assess mAb Glycosylation**
Karen Phinney - *NIST*

Karen
Phinney

Poster #6a : *Thursday : 11:00*   Glycomics

◎ **Lower CSF ApoE Glycosylation Associates with Measures of Alzheimer's Disease Biomarkers**
Dobrin Nedelkov - *Isoformix Inc*

Dobrin
Nedelkov

Poster #34a : *Thursday : 11:00* Glycomics

⊙ **One-pot Multienzyme (OPME) for Chemoenzymatic Generation of Affordable Gangliosides**
Andrew Lee - *IMCS*

Andrew Lee

Identifying High Value Tests

Poster #26b : *Thursday : 13:00* Identifying High Value Tests

⊙ **Antinuclear Antibodies (ANA) Platform Imperfections**
Mahesheema Ali - *The Metrohealth Medical Center*

Mahesheema Ali

Imaging

Poster #25b : *Wednesday : 13:00* Imaging

⊙ **A Novel Ion-beam Imaging Technique for Detection of Chromosomal Copy Number Alterations Associated with Progression of Breast Cancer**
Rashmi Kumar - *Stanford University*
Young Investigator Grantee

Rashmi Kumar

Poster #50b : *Thursday : 13:00* Imaging

⊙ **User-friendly Metabolic Segmentation and Anatomical Annotation of Mouse Brain Mass Spectrometry Images**
Ashley Woolfork - *University of Pennsylvania*

Ashley Woolfork

Lipidomics

Poster #33a : *Wednesday : 11:00* Lipidomics

⊙ **Odd Ones Out or In? Leveraging Odd-Chain Fatty Acids for Robust, Comprehensive Clinical Fatty Acid Analysis by Single Injection GC-MS**
Matthew Crawford - *Labcorp*

Matthew Crawford

Poster #5b : *Wednesday : 13:00* Lipidomics

⊙ **Development of Novel UHPLC-high Resolution Mass Spectrometry Approaches to Characterize Pancreatic Cancer Cells and Exosome Lipids**
Sina Feizbakhsh Bazargani - *University of Florida*

Sina Feizbakhsh Bazargani

Poster #17b : *Wednesday : 13:00*   Lipidomics

⊙ **The Development of a Liquid Chromatography – Mass Spectrometry Method for Investigating Lipid Metabolites in the Serum of Psoriatic Arthritis Patients**
John Koussiouris - *University of Toronto*
Young Investigator Grantee

John Koussiouris

Poster #10a : *Thursday : 11:00*   Lipidomics

⊙ **Comprehensive Lipidomic Profiling by Plasma Separation Cards**
Lauren Bishop - *University of California, Davis*

Lauren Bishop

Metabolomics

Poster #15b : *Wednesday : 13:00*   Metabolomics

⊙ **Simultaneous Detection of Salivary Cortisol and Cortisone Using an Automated High-throughput Sample Preparation Method for LC-MS/MS**
Ramisa Fariha - *Brown University*
Young Investigator Grantee

Ramisa Fariha

Poster #45b : *Wednesday : 13:00*   Metabolomics

⊙ **Comprehensive Targeted Metabolomics Method with Protocol for Reproducible HILIC Separation**
Karen Yannell - *Agilent Technologies*

Karen Yannell

Poster #6b : *Thursday : 13:00*   Metabolomics

⊙ **A Novel LC-MS Assay for the Quantification of Hydrogen Sulfide and Other Thiols in Biological Samples**
Hind Malaeb - *Cleveland State University*
Young Investigator Grantee

Hind Malaeb

Poster #38b : *Thursday : 13:00*   Metabolomics

⊙ **Enzymatic Methodology for the Selective Mass Spectrometric Investigation of Gut Microbiota-derived Metabolites**
Ioanna Tsiara - *Uppsala University*

Ioanna Tsiara

Poster #52b : *Thursday : 13:00*   Metabolomics

⊙ **Analysis of Acylcarnitines in Dried Blood Spot (DBS) Samples by FIA-MS/MS**
Jennifer King - *MilliporeSigma*

Jennifer King

Microbiology

Poster #35b : *Wednesday : 13:00*   Microbiology

◎ **Multi-Center Clinical Evaluation of Rapid Methicillin-resistant Staphylococcus aureus Screening Software based on MALDI-TOF MS**
Jong-Min Park - *Hallym University*
Young Investigator Grantee

Jong-Min Park

Multi-omics

Poster #1a : *Wednesday : 11:00*   Multi-omics

◎ **Biomarker Identification for Early Breast Cancer Diagnosis by Multi-omics Analysis**
Margret Thorsteinsdottir - *Faculty of Pharmaceutical Sciences, University of Iceland*

Margret Thorsteinsdottir

Practical Training

Poster #3a : *Wednesday : 11:00*   Practical Training

◎ **How Interpath has Transitioned into a Laboratory that Utilizes Mass Spectrometry for "High Complexity" Clinical Testing**
Peter Cohen - *Interpath Laboratory*

Peter Cohen

Pre-Analytics

Poster #18a : *Thursday : 11:00*   Pre-Analytics

◎ **In-Depth Analysis/Optimization of the ProTrap XG Sample Preparation Device Using Quantitative DIA Mass Spectrometry for Robust Plasma and Tissues Preparation**
Hugo Gagnon - *Phenoswitch Bioscience*

Hugo Gagnon

Poster #54a : *Thursday : 11:00*   Pre-Analytics

◎ **Effects of Blood Anticoagulants and Preanalytical Processing Delays on the Ex Vivo Proteome Content of Blood Platelets**
Joseph Aslan - *Knight Cardiovascular Institute, Oregon Health & Science University*

Joseph Aslan

Proteomics

Poster #7a : *Wednesday : 11:00*   Proteomics

◎ **Improved Quality Control for Swab Sampling for SARS-CoV-2 Detection by LC-MS/MS**
Wendy Heywood - *UCL Biological Mass Spectrometry Centre*

Wendy Heywood

Poster #13a : *Wednesday : 11:00*   Proteomics

○ **Highly Sensitive Quantification of Proteins from the SARS-CoV-2 Antigen in Nasopharyngeal Swab Samples**
Danielle Mackowsky - *Sciex*

Danielle Mackowsky

Poster #15a : *Wednesday : 11:00*   Proteomics

○ **Infliximab Quantitation by Tryptic Peptide LC-MS/MS; 10 Years of Method Improvements and Lessons Learned**
Paula Ladwig - *Mayo Clinic*

Paula Ladwig

Poster #21a : *Wednesday : 11:00*   Proteomics

○ **Teaching an Old Dog New Tricks: Expanding Targeted HDL-proteome Methods to Yield New Insights into Nonalcoholic Fatty Liver Disease Progression**
Timothy Collier - *Quest Diagnostics*

Timothy Collier

Poster #31a : *Wednesday : 11:00*   Proteomics

○ **Multiplexed Targeted Proteomics of Biomarkers for Lung Tumor Phenotyping and Treatment Selection**
John Koomen - *Moffitt Cancer Center*

John Koomen

Poster #49a : *Wednesday : 11:00*   Proteomics

○ **Gamma Heavy Chain Disease: Understanding the Structure of Truncated Immunoglobulin-G Heavy Chains Using MALDI-TOF, ESI-TOF, and Enzymatic Methods**
Ria Fyffe-Freil - *Mayo Clinic*

Ria Fyffe-Freil

Poster #55a : *Wednesday : 11:00*   Proteomics

○ **Optimization of Trypsin Digest for the Analysis of the SARS-CoV-2 Peptides**
Jennifer Kemp - *Mayo Clinic*

Jennifer Kemp

Poster #11b : *Wednesday : 13:00*   Proteomics

○ **Large Scale, Cloud-Enabled Re-Analysis of a Lung Cancer Cohort Yields Increased Plasma Proteome Depth and Putative Biomarkers**
Harendra Guturu - *Seer Inc*

Harendra Guturu

Poster #21b : *Wednesday : 13:00*   Proteomics

◎ **Immunoglobulin Relative Abundance in FFPE Tissue to Characterize Plasma Cell Neoplasms**
Jessica Chapman - *Memorial Sloan Kettering Cancer Center*

Jessica Chapman

Poster #47b : *Wednesday : 13:00*   Proteomics

◎ **Semi-automated and High-throughput Homogenisation Technique for In-depth Analysis of Various Tissue Proteome**
Zuzana Demianova - *PreOmics*

Zuzana Demianova

Poster #4a : *Thursday : 11:00*   Proteomics

◎ **Optimized Multi-Nanoparticle-based Plasma Proteomics with Enhanced Scale, Precision, and Depth of Coverage for Low Abundant Protein Biomarkers**
Daniel Hornburg - *Seer*

Daniel Hornburg

Poster #14a : *Thursday : 11:00*   Proteomics

◎ **Selective Separation and Sensitive Detection of Intact Parathyroid Hormone and its Variants by CESI-MS/MS**
Crystal Holt - *SCIEX*

Crystal Holt

Poster #22a : *Thursday : 11:00*   Proteomics

◎ **Mass Spectrometry-based Proteomics Biomarker Discovery and Verification in a Coronary Artery Disease Cohort**
Colleen Maxwell - *University of Leicester*
Young Investigator Grantee

Colleen Maxwell

Poster #24a : *Thursday : 11:00*   Proteomics

◎ **Deep Single-cell Type Proteome Profiling of Mouse Brain from Alzheimer's Disease Model by Nano-scale Tandem Mass Tag Mass Spectrometry**
Danting Liu - *St. Jude Children's Research Hospital*
Young Investigator Grantee

Danting Liu

Poster #56a : *Thursday : 11:00*   Proteomics

◎ **Quantitation of ODC in Brain Tumor Tissues**
Karthik Chandu - *Element Materials Technology*

Karthik Chandu

Poster #5a : *Wednesday : 11:00*   Tox / TDM / Endocrine

◎ **Reduced Hydrolysis Due to Natural Chemicals Inhibiting Glucuronidase Activity Resulting in Poor Liberation of Aglycones**
Anusha Chaparala - *Integrated Micro-Chromatography Systems Inc*

Anusha Chaparala

Poster #17a : *Wednesday : 11:00*   Tox / TDM / Endocrine

◎ **Simultaneous Quantitation of Immunosuppressant Drugs (cyclosporine A, Everolimus, Sirolimus, and Tacrolimus) in Human Whole Blood by LC-MS/MS**
Jessie Miao - *Milliporesigma*

Jessie Miao

Poster #19a : *Wednesday : 11:00*   Tox / TDM / Endocrine

◎ **Drink to That: A Novel SPE-Based Workflow for the Analysis of Phosphatidylethanols in Oral Fluid by LC/TQ**
Jennifer Hitchcock - *Agilent*

Jennifer Hitchcock

Poster #35a : *Wednesday : 11:00*   Tox / TDM / Endocrine

◎ **Detection of Fluorofentanyl in Chronic Pain and Behavioral Health Populations**
Lucas Marshall - *Aegis Sciences Corporation*

Lucas Marshall

Poster #41a : *Wednesday : 11:00*   Tox / TDM / Endocrine

◎ **Simultaneous Determination of Clofarabine, Fludarabine, Busulfan and Melphalan in Plasma by LC-MS/MS**
Ryan Schofield - *MSKCC*

Ryan Schofield

Poster #43a : *Wednesday : 11:00*   Tox / TDM / Endocrine

◎ **A Sensitive and Rapid UPLC-MS/MS Method for Simultaneous Quantitation of Five Newer Generation Antiepileptic Drugs and Two Active Metabolites in Human Serum**
Xiaowei Fu - *University of Tennessee Health Science Center*

Xiaowei Fu

Poster #45a : *Wednesday : 11:00*   Tox / TDM / Endocrine

◎ **The Analysis of 11 Steroids in Serum Using DPX µXTR Tips and LC-MS/MS**
Madison Kilpatrick - *DPX Labs*

Madison Kilpatrick

Poster #51a : *Wednesday : 11:00*   Tox / TDM / Endocrine

○ **Development of 13C-labeled Steroids for Use as Certified Reference Materials: The Journey to Aldosterone-13C3**
Heather Lima - *MilliporeSigma*

Heather Lima

Poster #53a : *Wednesday : 11:00*   Tox / TDM / Endocrine

○ **Quantification of Sirolimus and Everolimus in Whole Blood by a Rapid LC-MS/MS Method for Therapeutic Drug Monitoring**
Xiaoying Tang - *Cleveland Clinic*

Xiaoying Tang

Poster #3b : *Wednesday : 13:00*   Tox / TDM / Endocrine

○ **Highly Sensitive MS/MS Detection for Confident Identification of Potent Novel Synthetic Opioids and their Metabolites**
Pierre Negri - *SCIEX*

Pierre Negri

Poster #7b : *Wednesday : 13:00*   Tox / TDM / Endocrine

○ **Simultaneous Quantification of Infliximab and Adalimumab in Human Plasma by LC-MS/MS Using Ready-to-Use Kit and Comparison**
Mouton Nicolas - *Promise Proteomics*

Mouton Nicolas

Poster #9b : *Wednesday : 13:00*   Tox / TDM / Endocrine

○ **Rapid LC-MS/MS Method for Monitoring Bio-relevant Levels of Per- and Polyfluoroalkyl Substances (PFAS) in Serum**
Karl Oetjen - *SCIEX*

Karl Oetjen

Poster #19b : *Wednesday : 13:00*   Tox / TDM / Endocrine

○ **Bad Blood: Analysis of Phosphatidylethanols in Whole Blood by Novel Solid Phase Extraction and LC/TQ**
Tina Chambers - *Agilent*

Tina Chambers

Poster #23b : *Wednesday : 13:00*   Tox / TDM / Endocrine

○ **Cross-validation of a Multiplex LC-MS/MS Method for Assaying mAbs Plasma Levels in Patients with Cancer**
Dorothee Lebert - *Promise Proteomics*

Dorothee Lebert

Poster #31b : *Wednesday : 13:00*   Tox / TDM / Endocrine

○ **Analytical Performance Characteristics of a Definitive UPLC-MS/MS Assay for Quantitative Measurement of 11 Opioid Compounds in Urine**
Robert Maynard - *University of North Carolina*

Robert Maynard

Poster #33b : *Wednesday : 13:00*　　　　　　　　　　　Tox / TDM / Endocrine

◎ Improving Imprecisely Imperfect Immunosuppressant's; Foundations that Enhance Methodological Accuracy and Imprecision
Russell Grant - *Labcorp*

Russell Grant

Poster #41b : *Wednesday : 13:00*　　　　　　　　　　　Tox / TDM / Endocrine

◎ Targeted Forensic Screening and Semi-quantitation of Drugs in Urine Using a Novel High-resolution Accurate-mass Mass Spectrometer
Kristine van Natta - *Thermo Fisher Scientific*

Kristine van Natta

Poster #49b : *Wednesday : 13:00*　　　　　　　　　　　Tox / TDM / Endocrine

◎ In-Utero Drug Exposure Rates in Neonates Before and During the Covid-19 Pandemic: A Retrospective Analysis of Positive Drug Detection in Meconium
Brooke Andrews - *University of Kentucky*

Brooke Andrews

Poster #51b : *Wednesday : 13:00*　　　　　　　　　　　Tox / TDM / Endocrine

◎ Confidence in Your Calibrators: Metrologically Traceable Calibrators and Quality Controls for the LC-MS Analysis of Steroid Hormones
Joseph Clarke - *Waters Corporation*

Joseph Clarke

Poster #12a : *Thursday : 11:00*　　　　　　　　　　　Tox / TDM / Endocrine

◎ Quantification of Thiopurine Metabolites and Correlation with TPMT Phenotype
Amol Bajaj - *ARUP*

Amol Bajaj

Poster #16a : *Thursday : 11:00*　　　　　　　　　　　Tox / TDM / Endocrine

◎ High-Throughput Analysis of 34 Drugs of Abuse in Human Urine Using LC-MS/MS
Xu Zhang - *Alberta Centre for Toxicology*

Xu Zhang

Poster #20a : *Thursday : 11:00*　　　　　　　　　　　Tox / TDM / Endocrine

◎ Quantitative LC-MS/MS Analysis of Ethyl Glucuronide and Ethyl Sulfate in Urine Using Poroshell Charged Surface C18 Column
Natalie Rasmussen - *Agilent Technologies*

Natalie Rasmussen

Poster #26a : *Thursday : 11:00*　　　　　　　　　　　Tox / TDM / Endocrine

◎ Utility of Detecting Fentanyl Analogs During LC-MS/MS Confirmation for Positive Fentanyl Urine Drug Screens
Catherine Omosule - *Washington Univ in St. Louis*
Young Investigator Grantee

Catherine Omosule

Poster #30a : *Thursday : 11:00*   Tox / TDM / Endocrine

◎ Quantification of Urinary Fractionated Metanephrines
Using Solid Phase Extraction and LC-MS/MS
Adam Kutnick - *Cleveland Clinic*

Adam
Kutnick

Poster #32a : *Thursday : 11:00*   Tox / TDM / Endocrine

◎ New Software Tool for a Routine High-
resolution LC/Q-TOF Screening Workflow for Drug
Analysis
Cate Simmermaker - *Agilent Technologies*

Cate
Simmermaker

Poster #38a : *Thursday : 11:00*   Tox / TDM / Endocrine

◎ Integration of Verispray and FAIMS for Quantitation of
Immunosuppressant Drugs in Whole Blood
Kyana Garza - *Johns Hopkins University School of Medicine*

Kyana
Garza

Poster #46a : *Thursday : 11:00*   Tox / TDM / Endocrine

◎ Toxicology Screening of Human Blood Using
Quadrupole-Time of Flight (QTOF) Mass
Spectrometry
Evelyn Wang - *Shimadzu Scientific Instruments*

Evelyn
Wang

Poster #48a : *Thursday : 11:00*   Tox / TDM / Endocrine

◎ Optimization of a Rapid Qualitative Drug Screening
Platform Using Coated Blade Spray – Mass Spectrometry
(CBS-MS/MS)
Folagbayi Arowolo - *UCSF*

Folagbayi
Arowolo

Poster #50a : *Thursday : 11:00*   Tox / TDM / Endocrine

◎ Semi-Quantitative Analysis of 49 Drugs of Abuse
from Meconium by LC-MS/MS
Triniti Jensen - *ARUP*

Triniti
Jensen

Poster #52a : *Thursday : 11:00*   Tox / TDM / Endocrine

◎ High-throughput, Sensitive and Selective LC-MS/MS
Method for Quantitation of Serum Estradiol Utilizing Fmp-ts
Derivatization
Caroline Wentworth - *ARUP*

Caroline
Wentworth

Poster #8b : *Thursday : 13:00*   Tox / TDM / Endocrine

◎ Low Level Quantification of Eight Vitamin D
Metabolites by UPLC-MS/MS for Clinical Research
Peter Harrsch - *Waters Corporation*

Peter
Harrsch

Poster #10b : *Thursday : 13:00*   Tox / TDM / Endocrine

○ **Use of QTOF to Detect Novel Psychoactive Substances in Patients Receiving Treatment for Substance Use Disorder**
David Newcombe - *Ideal Option*

David Newcombe

Poster #14b : *Thursday : 13:00*   Tox / TDM / Endocrine

○ **Improved Sensitivity for Aldosterone Using the Unique MRM3 Quantification Workflow**
Xiang He - *SCIEX*

Xiang He

Poster #20b : *Thursday : 13:00*   Tox / TDM / Endocrine

○ **Plasma Catecholamine Quantification by Reverse Phase LC-MS/MS with In-well Ion Pairing**
Stephen Merrigan - *ARUP Laboratories*

Stephen Merrigan

Poster #30b : *Thursday : 13:00*   Tox / TDM / Endocrine

○ **Improving Drug Analysis from a Single Fingerprint for Medical Adherence Monitoring**
Katie Longman - *University of Surrey*

Katie Longman

Poster #32b : *Thursday : 13:00*   Tox / TDM / Endocrine

○ **Breaking Down the Barrier to Rapid Antibiotic Levels with Paper Spray-Mass Spectrometry (PS-MS/MS): Simultaneous Quantitation of Five β-lactams from Plasma**
Lindsey Kirkpatrick - *Indiana University School of Medicine*

Lindsey Kirkpatrick

Poster #36b : *Thursday : 13:00*   Tox / TDM / Endocrine

○ **Evaluation of New beta-glucuronidase for Urine Drug Testing**
Rasha Abddelgader - *Baylor Scott and White Health*

Rasha Abddelgader

Poster #42b : *Thursday : 13:00*   Tox / TDM / Endocrine

○ **Immunoassay vs. LC-MS/MS, Mitragynine**
Lei Shao - *Thermo Fisher Scientific*

Lei Shao

Poster #44b : *Thursday : 13:00*   Tox / TDM / Endocrine

○ **Consolidating LC-MS/MS Method Conditions for the Analysis of Alcohol Metabolites, Barbiturates, and Drugs of Abuse**
Jonathan Edelman - *Restek*

Jonathan Edelman

Poster #48b : *Thursday : 13:00*   Tox / TDM / Endocrine

○ **Antifungal drug monitoring assay validation on LC-MS/MS**
Mark Girton - *University of Virginia*

Mark
Girton

Troubleshooting

Poster #28a : *Tuesday : 20:00*   Troubleshooting

○ **A Stumbling Block of Harmonizing LC-MS/MS Assays in Clinical Laboratories**
Hsuan-Chieh (Joyce) Liao - *University of Washington*

Hsuan-Chieh (Joyce)
Liao

Poster #28b : *Tuesday : 20:15*   Troubleshooting

○ **Morphine and Oxycodone Co-Positivity in Pain Management Urine Drug Testing**
Stephen Roper - *Washington University School of Medicine*

Stephen
Roper

Poster #27a : *Tuesday : 20:30*   Troubleshooting

○ **Nonspecific Adsorption and Loss of 11-Nor-9-carboxy-Δ9-tetrahydrocannabinol. Dude, where's my THC?**
Triniti Jensen - *ARUP*

Triniti
Jensen

Poster #27b : *Tuesday : 20:45*   Troubleshooting

○ **Interfering Peak in the Estradiol (E2U) LC-MS/MS Assay**
Mima Geere - *UCSF*

Mima
Geere

Poster #29a : *Wednesday : 17:45*   Troubleshooting

○ **Interfering Compound in Urine Cannabinoid Analysis**
Agnes Cua - *Precision Diagnostics*

Agnes
Cua

Poster #29b : *Wednesday : 18:00*   Troubleshooting

○ **Communicating Proteogenomics Data to Physicians**
John Koomen - *Moffitt Cancer Center*

John
Koomen

Poster #9a : *Wednesday : 11:00*

◉ **Pharmacokinetic and Brain Distribution Study of an Anti-Glioblastoma Agent in Mice by HPLC-MS/MS**
Yaxin Li - *Cleveland State University*

Yaxin
Li

Poster #55b : *Wednesday : 13:00*

◉ **The Impact of 25-hydroxyvitamin D Levels on COVID-19 Severity**
Nguyen Nguyen - *Baylor Scott & White Health*

Nguyen
Nguyen

Poster #2b : *Thursday : 13:00*

◉ **Quantitation of Clinical Research Pain Panel Analytes from Oral Fluid Utilizing Microelution Solid Phase Extraction Coupled with LC-MS/MS**
Shahana Huq - *Phenomenex*

Shahana
Huq

Poster #54b : *Thursday : 13:00*

◉ **A Single Pipette Tip Washing Protocol is Sufficient to Reduce Carryover in a Panel of 53 Drug Analytes to Undetectable Levels via ELISA**
Carter Swanson - *Grenova*

Carter
Swanson

My Poster Viewing Plan

☐ *Wed : 13:00 :* **#11b** Large Scale, Cloud-Enabled Re-An...	**Guturu**	Proteomics
☐ *Wed : 13:00 :* **#13b** New NIST Reference Material Help...	**Phinney**	Glycomics
☐ *Wed : 13:00 :* **#15b** Simultaneous Detection of Saliva...	**Fariha**	Metabolomics
☐ *Wed : 13:00 :* **#17b** The Development of a Liquid Chro...	**Koussio...**	Lipidomics
☐ *Wed : 13:00 :* **#19b** Bad Blood: Analysis of Phosphat...	**Chambers**	Tox / TDM / ...
☐ *Wed : 13:00 :* **#21b** Immunoglobulin Relative Abundanc...	**Chapman**	Proteomics
☐ *Wed : 13:00 :* **#23b** Cross-validation of a Multiplex ...	**Lebert**	Tox / TDM / ...
☐ *Wed : 13:00 :* **#25b** A Novel Ion-beam Imaging Techniq...	**Kumar**	Imaging
☐ *Wed : 13:00 :* **#31b** Analytical Performance Character...	**Maynard**	Tox / TDM / ...
☐ *Wed : 13:00 :* **#33b** Improving Imprecisely Imperfect ...	**Grant**	Tox / TDM / ...
☐ *Wed : 13:00 :* **#35b** Multi-Center Clinical Evaluation...	**Park**	Microbiology
☐ *Wed : 13:00 :* **#37b** Ion Ratio Evaluation Using a Lin...	**Koch**	Data Analytics
☐ *Wed : 13:00 :* **#39b** A Peptide Enrichment and Mass Sp...	**Gibson**	Assays Lever...
☐ *Wed : 13:00 :* **#41b** Targeted Forensic Screening and ...	**van Natta**	Tox / TDM / ...
☐ *Wed : 13:00 :* **#43b** Automating Low-Volume 96-Well SP...	**Senior**	Emerging Tec...
☐ *Wed : 13:00 :* **#45b** Comprehensive Targeted Metabolom...	**Yannell**	Metabolomics
☐ *Wed : 13:00 :* **#47b** Semi-automated and High-throughp...	**Demianova**	Proteomics
☐ *Wed : 13:00 :* **#49b** In-Utero Drug Exposure Rates in ...	**Andrews**	Tox / TDM / ...
☐ *Wed : 13:00 :* **#51b** Confidence in Your Calibrators: ...	**Clarke**	Tox / TDM / ...
☐ *Wed : 13:00 :* **#53b** Better Together: A Novel Ion-pai...	**Szczesn...**	Assays Lever...
☐ *Wed : 13:00 :* **#55b** The Impact of 25-hydroxyvitamin ...	**Nguyen**	Various OTHER

Troubleshooting Poster Session B

☐ *Wed : 17:45 :* **#29a** Interfering Compound in Urine Ca...	**Cua**	Troubleshooting
☐ *Wed : 18:00 :* **#29b** Communicating Proteogenomics Dat...	**Koomen**	Troubleshooting

Poster Session 3

☐ *Thu : 11:00 :* **#2a** Quantitative Analysis of Vitamin...	**Vachali**	Assays Lever...
☐ *Thu : 11:00 :* **#4a** Optimized Multi-Nanoparticle-bas...	**Hornburg**	Proteomics
☐ *Thu : 11:00 :* **#6a** Lower CSF ApoE Glycosylation Ass...	**Nedelkov**	Glycomics
☐ *Thu : 11:00 :* **#10a** Comprehensive Lipidomic Profilin...	**Bishop**	Lipidomics
☐ *Thu : 11:00 :* **#12a** Quantification of Thiopurine Met...	**Bajaj**	Tox / TDM / ...
☐ *Thu : 11:00 :* **#14a** Selective Separation and Sensiti...	**Holt**	Proteomics
☐ *Thu : 11:00 :* **#16a** High-Throughput Analysis of 34 D...	**Zhang**	Tox / TDM / ...
☐ *Thu : 11:00 :* **#18a** In-Depth Analysis/Optimization o...	**Gagnon**	Pre-Analytics
☐ *Thu : 11:00 :* **#20a** Quantitative LC-MS/MS Analysis o...	**Rasmussen**	Tox / TDM / ...
☐ *Thu : 11:00 :* **#22a** Mass Spectrometry-based Proteomi...	**Maxwell**	Proteomics
☐ *Thu : 11:00 :* **#24a** Deep Single-cell Type Proteome P...	**Liu**	Proteomics
☐ *Thu : 11:00 :* **#26a** Utility of Detecting Fentanyl An...	**Omosule**	Tox / TDM / ...
☐ *Thu : 11:00 :* **#30a** Quantification of Urinary Fracti...	**Kutnick**	Tox / TDM / ...
☐ *Thu : 11:00 :* **#32a** New Software Tool for a Routine ...	**Simmerm...**	Tox / TDM / ...
☐ *Thu : 11:00 :* **#34a** One-pot Multienzyme (OPME) for C...	**Lee**	Glycomics

☐ *Thu : 11:00 :* **#36a**	Middle-Up Approach for Therapeut...	**Song**	Assays Lever...
☐ *Thu : 11:00 :* **#38a**	Integration of Verispray and FAI...	**Garza**	Tox / TDM / ...
☐ *Thu : 11:00 :* **#40a**	Direct Quantitation of Phosphati...	**Guo**	Emerging Tec...
☐ *Thu : 11:00 :* **#42a**	Shaping the Future of Estrogen A...	**Weatherill**	Emerging Tec...
☐ *Thu : 11:00 :* **#46a**	Toxicology Screening of Human Bl...	**Wang**	Tox / TDM / ...
☐ *Thu : 11:00 :* **#48a**	Optimization of a Rapid Qualitat...	**Arowolo**	Tox / TDM / ...
☐ *Thu : 11:00 :* **#50a**	Semi-Quantitative Analysis of 49...	**Jensen**	Tox / TDM / ...
☐ *Thu : 11:00 :* **#52a**	High-throughput, Sensitive and S...	**Wentworth**	Tox / TDM / ...
☐ *Thu : 11:00 :* **#54a**	Effects of Blood Anticoagulants ...	**Aslan**	Pre-Analytics
☐ *Thu : 11:00 :* **#56a**	Quantitation of ODC in Brain Tum...	**Chandu**	Proteomics

Poster Session 4

☐ *Thu : 13:00 :* **#2b**	Quantitation of Clinical Researc...	**Huq**	Various OTHER
☐ *Thu : 13:00 :* **#6b**	A Novel LC-MS Assay for the Quan...	**Malaeb**	Metabolomics
☐ *Thu : 13:00 :* **#8b**	Low Level Quantification of Eigh...	**Harrsch**	Tox / TDM / ...
☐ *Thu : 13:00 :* **#10b**	Use of QTOF to Detect Novel Psyc...	**Newcombe**	Tox / TDM / ...
☐ *Thu : 13:00 :* **#12b**	Analysis of Estrogens in Plasma ...	**Chen**	Assays Lever...
☐ *Thu : 13:00 :* **#14b**	Improved Sensitivity for Aldoste...	**He**	Tox / TDM / ...
☐ *Thu : 13:00 :* **#16b**	Utilizing LC-MS/MS for the Clini...	**Hauck**	Assays Lever...
☐ *Thu : 13:00 :* **#18b**	Small Footprint, Big Screen: Dev...	**Smith**	Assays Lever...
☐ *Thu : 13:00 :* **#20b**	Plasma Catecholamine Quantificat...	**Merrigan**	Tox / TDM / ...
☐ *Thu : 13:00 :* **#22b**	Detection of Metabolic Urine Org...	**Ganepola**	Assays Lever...
☐ *Thu : 13:00 :* **#24b**	Analytical Validation of a Glyco...	**Wong**	Assays Lever...
☐ *Thu : 13:00 :* **#26b**	Antinuclear Antibodies (ANA) Pla...	**Ali**	Identifying ...
☐ *Thu : 13:00 :* **#30b**	Improving Drug Analysis from a S...	**Longman**	Tox / TDM / ...
☐ *Thu : 13:00 :* **#32b**	Breaking Down the Barrier to Rap...	**Kirkpat...**	Tox / TDM / ...
☐ *Thu : 13:00 :* **#34b**	A Story of Discovery and Transla...	**Kuzdzal**	Assays Lever...
☐ *Thu : 13:00 :* **#36b**	Evaluation of New beta-glucuroni...	**Abddelg...**	Tox / TDM / ...
☐ *Thu : 13:00 :* **#38b**	Enzymatic Methodology for the Se...	**Tsiara**	Metabolomics
☐ *Thu : 13:00 :* **#40b**	Simultaneous Determination of Tr...	**Balcı**	Assays Lever...
☐ *Thu : 13:00 :* **#42b**	Immunoassay vs. LC-MS/MS, Mitrag...	Shao	Tox / TDM / ...
☐ *Thu : 13:00 :* **#44b**	Consolidating LC-MS/MS Method Co...	**Edelman**	Tox / TDM / ...
☐ *Thu : 13:00 :* **#46b**	Development and Validation of a ...	**Olayinka**	Assays Lever...
☐ *Thu : 13:00 :* **#48b**	Antifungal drug monitoring assay...	**Girton**	Tox / TDM / ...
☐ *Thu : 13:00 :* **#50b**	User-friendly Metabolic Segmenta...	**Woolfork**	Imaging
☐ *Thu : 13:00 :* **#52b**	Analysis of Acylcarnitines in Dr...	**King**	Metabolomics
☐ *Thu : 13:00 :* **#54b**	A Single Pipette Tip Washing Pro...	**Swanson**	Various OTHER

Exhibitors

Agilent Technologies (Booth #9-10)
https://www.agilent.com/
Agilent Technologies delivers premiere analytical technologies for clinical research ensuring your success from sample prep to final answer. These include a comprehensive portfolio of innovative automation, chemistries, GC, GC/MS, ICP/MS, LC, and LC/MS solutions which enables the identification and quantification of both endogenous and exogenous substances in complex biological matrices with the utmost accuracy and reliability. Coupled with our dedicated global support network, we will get you to your final answer with minimal ramp-up and maximum productivity.

Analytik Jena (Booth #20)
https://www.analytik-jena.us/
Analytik Jena is a leading provider of high-end analytical measuring technology, of instruments and products in the fields of biotechnology and molecular diagnostics, as well as of high quality liquid handling and automation technologies. Its portfolio includes traditional analytical technology, particularly to measure concentrations of elements and molecules, as well as systems for bioanalytical applications in the Life Science area spanning the highly complex analytic cycle of a sample from sample preparation to detection. Automated high-throughput screening systems for the pharmaceutical sector are also part of this segment's extensive portfolio. Analytik Jena´s products are focused to offer customers and users a quality and the reproducibility of their laboratory results. Services, as well as device-specific consumables and disposables, such as reagents or plastic articles, complete the Group's extensive range of products.

BaySpec (Booth #11)
https://www.bayspec.com/
BaySpec, Inc., founded in 1999 with 100% manufacturing in the USA (San Jose, California), is a vertically integrated spectral sensing company. The company designs, manufactures and markets advanced spectral instruments, including UV-VIS-NIR-SWIR spectrometers, benchtop and portable NIR/SWIR and Raman analyzers, confocal Raman microscopes, hyperspectral imagers, mass spectrometers, and OEM spectral engines and components, for the R&D, biomedical, pharmaceuticals, chemical, food, semiconductor, health monitoring, human & animal medical devices, and the optical telecommunications industries.

Biotage (Booth #30)
https://www.biotage.com
Biotage® is a leading provider of sample preparation instrumentation and consumables for a wide range of applications, including Clinical, Pharmaceutical, and Forensic applications. ISOLUTE® and EVOLUTE® brand solid-phase extraction (SPE) and Supported Liquid Extraction (SLE) products can be run in either a manual or automated environment. The Biotage® Extrahera™ Automated workstation and TurboVap® Solvent Evaporators are ideal for increasing throughput and achieving accurate results. Stop by our booth to view the latest innovations in Evaporation and Sample preparation - the new Biotage® Extrahera LV-200 workstation built to process low volume samples with ease and our latest evaporation system the Biotage® TurboVap 96 Dual that allows the use of two single evaporators and a dual system all within the same footprint!

Chrom Tech (Mini-Table)
http://www.chromtech.com
Distributor of Chromatography consumables, instrumentation and supplies. Featuring: Sample Preparation Products, 96 Well Plates for MS, 96-well Multi-Tier™ Micro Plate System with Glass Inserts, Columns, Instrument consumables and replacement parts, Pumps, Gas Generators. Featured Suppliers include: Agilent Technologies, Thermo Scientific, Sigma Aldrich, Idex (Upchurch and Rheodyne), Parker Balston, Hamilton, Restek.

Chromsystems (Booth #29)

http://www.chromsystems.com

Chromsystems is a leading global company providing ready-to-use kits, multilevel calibrators and quality controls for routine clinical diagnostics by LC-MS/MS and HPLC. Our parameter menu covers a range of areas such as newborn screening, therapeutic drug monitoring, steroid analysis, vitamin profiling and more. We continuously expand our portfolio with additional tests all ensuring a highly accurate and cost-effective analysis. We enable laboratories to add new parameters into their diagnostic routine and expand their testing menu without prior technical expertise. They can immediately start the analysis with a minimum of time for the sample preparation. The products are comprehensively validated, and in particular LC-MS/MS methods with all widely used tandem mass spectrometers. They are CE-IVD compliant, satisfying regulatory requirements in the laboratory. We combine these high quality products with an excellent support programme and service for our customers.

DPX Technologies (Booth #23)

http://www.dpxtechnologies.com/

At DPX Technologies, we believe that sample preparation should be fast and simple. That is why we have developed a variety of sample preparation product lines in easy-to-use pipette tips. Our patented dispersive pipette eXTRaction (XTR) tip functions by dispersive SPE, requiring only seconds of mixing to complete each step of the sample preparation process. The patented Tip-on-Tip technology employs a DPX's Filtration tips to filter samples by dispensing them through the Filtration tip. The latest DPX product line, INTip Size Exclusion Chromatography, provides a unique, patent pending process for automated group separations for buffer exchange, desalting and PCR cleanup. Whether your laboratory uses a single channel pipettor or a fully robotic liquid handler, there is a DPX tip compatible with your analysis method and throughput. Stop by our booth to learn how DPX products can streamline your workflow.

Evosep (Mini-Table)

https://www.evosep.com/

Evosep aims to improve quality of life and patient care by radically innovating protein based clinical diagnostics. Making sample preparation and separation before MS analysis 10 times faster and 100 times more robust will enable truly large cohort studies for biomarker validation and provide the foundation for precision medicine.

Golden West Diagnostics (Booth #40)

http://www.goldenwestdiagnostics.com

Golden West Diagnostics, Inc. addresses the need for quality, cost effective biological raw materials for the development of immunoassays and LC-MS applications. GWD provides manufacturers and laboratories with over 80 products including Vitamin D free human serum, serum for ultra-sensitive testing, HSA, HGG, and RGG. Please visit us at www.goldenwestdiagnostics.com.

GRENOVA (Booth #24)

http://www.grenovasolutions.com

GRENOVA offers laboratories an innovative option for plastic pipette tip re-use. Our patented technology allows for both manual and automated pipette tip washing, sterilizing, drying, and storage. GRENOVA devices can save up to 90% on your consumables and bio-waste spend, as well as support your green initiatives by reducing waste stream pollution.

Hamilton Company (Booth #12)

https://www.hamiltoncompany.com/

For over 60 years, trust has been the hallmark of every precision-crafted solution at Hamilton Company. Private, public, and academic institutions across the globe and throughout diverse scientific and clinical reams trust in Hamilton's ergonomic manual precision measurement devices and powerful automated liquid handling workstations to streamline cumbersome workflows and aid in answering complex questions.

The world's top equipment manufacturers trust in Hamilton's high-quality solutions and expertise to develop robust, cutting-edge solutions in less time and with lower development and manufacturing costs than ever before. These solutions, both large and small, can help companies to make an impact with their target customers, strengthen the path to success, and stand apart from the competition.

HORIZON Lab Systems (Booth #33)
http://www.horizonlims.com
HORIZON is today's approach to lab information management, based on more than 30 years of industry leadership. Originally developed by ChemWare, HORIZON Lab Systems LLC is a part of the Dohmen family of companies. Some of the most mission-critical marketplaces apply our product – each one important to millions every day. From large government public health labs to small, private clinical ones, HORIZON is designed to be the LIMS of choice for any lab. If you collect, process and test samples, HORIZON is your lab's solution.

IMCS (Booth #32)
http://www.imcstips.com
Integrated Micro-Chromatography Systems, Inc (IMCS) designs, develops, manufactures, and distributes next-generation recombinant proteins and chromatography consumables to clinical and forensic toxicology laboratories, academic research facilities, US government agencies, and life science companies worldwide. The company has two product lines: IMCSzyme® and IMCStips®. The industry-disrupting IMCSzyme is a genetically modified enzyme primarily used for the drug screening and analysis of urine and other biological samples. IMCStips are utilize patented dispersive solid-phase extraction (dSPE) technology for purifying new proteins, antibodies, and enzymes, as part of drug discovery from validation to development, screening, and manufacturing. Recently, IMCS was awarded NIH funds to manufacture various glycosyltransferases to leverage the company's advanced manufacturing and research capabilities to expand the synthesis of glycosphingolipids and sialoglycans. This expands IMCS product lines and delivers critically needed biological reagents that were previously unavailable to the scientific community.

Immundiagnostik (Mini-Table)
http://www.immundiagnostik.com
Immundiagnostik AG (www.immundiagnostik.com), founded in 1986 by Dr. Franz Paul Armbruster (CEO), is specialized on the development, production, and world-wide distribution of innovative parameters and detection methods for laboratory diagnostics and medical research. The main focus is the development of immunological tools, of HPLC and molecular biology methods, and of new applications for mass spectrometry (LC-MS/MS). Immundiagnostik concentrates on the development and production of laboratory diagnostics for the identification of disease risks, for differential diagnosis, and for therapeutic drug monitoring. The company holds a particularly strong portfolio in markers of oxidative stress/anti-aging, gastroenterology and nutrition, skeletal system, and cardio-reno-vascular system. Immundiagnostik owns more than 35 patents in Europe, the US, Japan, Canada, and Australia, is certified according to DIN EN ISO 13485 and fulfills the requirements of the German Medical Device Regulation and the EU IVD Regulations (98/79 EG).

Indigo BioAutomation (Booth #21-22)
http://www.indigobio.com/
Indigo BioAutomation, founded in 2004, is an established leader in software automation for the applied and health sciences. Indigo's flagship system, ASCENT, is a hosted system for automating the processing, reviewing, and reporting of LC-MS/MS data. ASCENT has helped automate the review of tens of millions of samples, in a variety of clinical/toxicology laboratories. In addition to daily workflows, Indigo's systems also provide laboratory analytics dashboards. Looking ahead, Indigo will continue to create world-class computational decision-making tools and laboratory automation systems.

Jasem (Booth #19)

http://www.jasem.com.tr

Adapting diagnostics in chromatography coupled mass spectrometry Jasem serves customers ready-to-use diagnostic kits for clinical analysis based on HPLC and mass spectrometry. Providing innovative trustworthy and accurate results constitutes the core viewpoint of us. Hence, our straightforward and economic solutions are being used extensively in clinical and food laboratories. Foundation purpose is clinical and food analysis kit development; Simple and practical sample preparation Short analysis time Reliable and sensitive analysis Low analysis cost (without derivatization, SPE or concentration) Longer column life-time

Kura Biotech (Booth #15)

http://www.kurabiotech.com/

We create biotech products better and differently to be a force of good to the world. Kura Biotech is a Chilean biotechnology company that, motivated by generating a positive impact in the world and driven by a limitless spirit, has become the world's largest producer of enzymatic reagents for drug detection. Now we are adopting that expertise in proteomics analysis by developing superior enzymatic tools to streamline sample preparation workflows enabling scientists to overcome challenges all over the world. Don't miss Finden's brand B-One enzyme, the best and most reliable beta-glucuronidase in the forensic and clinical toxicology market. It allows incredibly fast, clean, and convenient glucuronide hydrolysis at room-temperature for high-throughput compatibility. An All-In-One reagent, stabilized in reaction buffer and stable for 12-weeks at room-temperature. Simply combine the urine sample, internal standards, B-One, mix and inject. Easy as 1-2-3!

Mass Tech (Booth #38)

http://www.apmaldi.com

MassTech is the only license holder and US producer of AP MALDI and subAP MALDI ion sources and is the recognized leader in AP MALDI technology. AP MALDI is suitable for sensitive MALDI sample analysis and below 10-micron spatial resolution MS tissue imaging. The AP MALDI sources can be changed to the ESI source within minutes. AP/MALDI is compatible with HRMS from all MS OEMs and select triple quadrupole MS systems. The MT Explorer 30 is a compact mass spectrometer designed and built by MassTech Inc. It is capable of desktop laboratory performance in a small, field transportable package. It has an atmospheric pressure interface, allowing for a use of ESI/DART/APCI/MALDI/SICRIT ion sources. Combined with our own ESI/APCI ion source, customers can purchase a complete integrated unit capable of rapid, on-site biological or chemical identification.

MilliporeSigma (Booth #36-37)

http://www.sigmaaldrich.com

MilliporeSigma, the life science business of Merck KGaA, Darmstadt, Germany, is a leading company for innovative and top-quality high-tech products in healthcare, life science and performance materials. MilliporeSigma has created the Supelco® portfolio of analytical products, putting quality and compliance at the forefront of our work to ensure your results are reproducible and your systems fully certified. Our robust portfolio, developed by analytical chemists for analytical chemists, covers a broad range of analytical solutions and every product undergoes meticulous quality control to maintain the integrity of your testing protocols to provide accuracy and reliability, every time.

Newomics (Booth #28)

https://www.newomics.com/

Newomics Inc. is dedicated to creating and commercializing innovative (New) and integrative (Omics) platforms and solutions for precision medicine. Our initial M3 emitter product line enables unprecedented gains in sensitivity, throughput, and robustness for liquid chromatography – mass spectrometry (LC-MS). The multinozzle emitter array (MEA) technology, invented at the Lawrence Berkeley National Laboratory, won the 2012 R&D 100 award. We are applying our technology to develop multiomics diagnostic assays for diabetes, Alzheimer's, and other diseases. We are also developing a microfluidic device to sort, count, and remove cells based on size from blood (the Senescence Chip). The impact of our publications led to a ranking of #71 among the top 100 corporations in the US that dominated research in the natural sciences in 2018.

Optimize Technologies (Booth #31)

http://www.optimizetech.com

Optimize Technologies offers a complete line of innovative components and replacement parts for UHPLC, HPLC and LC/MS systems. Products include EXP® Fittings, Filters, Traps and Guards, OPTI-MAX® Check Valves, OPTI-SEAL® Seals, Replacement Pistons, OPTI-GUARD® Guard Columns, OPTI-PAK® Traps, OPTI-SOLV® Filters and OPTI-LYNX™ Quick-Connect packed beds. New products include EXP® hand-tight fittings, UHPLC/MS traps, UHPLC filtration, guard solutions rated to 20,000+ psi and OPTI-TRAPS™ for large molecules, peptides, online desalting and detergent removal. All Optimize EXP® products feature hand-tight holders and EXP® Titanium Hybrid reusable ferrules.

PerkinElmer (Booth #4)

https://www.perkinelmer.com

PerkinElmer, Inc. is a global leader focused on improving the health and safety of people and their environment. PerkinElmer is dedicated to the quality and sustainability of the environment. With our analytical instrumentation, illumination and detection technologies, and leading laboratory services, we focus on improving the integrity and safety of the world we live in.

Phenomenex (Booth #3)

http://www.phenomenex.com

Phenomenex is a global technology leader committed to developing novel analytical chemistry solutions that solve the separation and purification challenges of researchers in industrial, government and academic laboratories. Phenomenex's core technologies include products for liquid chromatography, gas chromatography, sample preparation, bulk purification chromatographic media, and chromatography accessories and equipment. For more information, visit www.phenomenex.com.

PreOmics (Booth #25)

http://www.preomics.com

PreOmics – Setting the Standard for Protein Analysis. PreOmics empowers our clients in life sciences to establish biological knowledge through efficient, reliable solutions and workflows that set the standard for protein analysis. Our team spirit, energy, and commitment empower us to be both creative and quality focused – A trusted partner with deeply rooted scientific experience. We envisage a future with revolutionary proteomic discovery processes open for everyone. Tools that reveal hidden causes of diseases, ensure sustainable nutrition, and provide diagnoses that enhance lives and society.

Proteoform Scientific (Mini-Table)

https://proteoform.com/

We are dedicated to accelerating the pace of biotech and biopharma discoveries. Proteoform Scientific recently joined forces with CRO, PhenoSwitch Bioscience to operate as a cohesive unit delivering multiomics sample prep, mass spec services, and data analytics solutions to our customers. Proteoform's patented proteomics sample preparation technology along with PhenoSwitch's collaborative approach to experiment design and delivery and reaching new levels of clarity in data and research results help customers speed up the development path from the lab bench to patient bedside.

Restek (Booth #35)

http://www.restek.com

A leading innovator of chromatography solutions for both LC and GC, Restek has been developing and manufacturing columns, reference standards, sample preparation materials, accessories, and more since 1985. We provide analysts around the world with products and services to monitor the quality of air, water, soil, food, pharmaceuticals, chemicals, and petroleum products. Our experts have diverse areas of specialization in chemistry, chromatography, engineering, and related fields as well as close relationships with government agencies, international regulators, academia, and instrument manufacturers. www.restek.com

SCIEX (Booth #1,2,13,14)
https://sciex.com/diagnostics
At SCIEX, our mission is to deliver solutions for the precision detection and quantification of molecules, empowering our customers to protect and advance the wellness and safety of all. SCIEX has led the field of mass spectrometry for 50 years. From the moment we launched the first ever commercially successful triple quad in 1981, we have developed groundbreaking technologies and solutions that influence life-changing research and outcomes. Today, as part of the Danaher family of global life science and technology innovators, we continue to pioneer robust solutions in mass spectrometry and capillary electrophoresis. But we don't just develop products. It is what we do together with our customers that we are most proud of. That's why thousands of experts around the world choose SCIEX to get the answers they can trust.

Shimadzu (Booth #26)
http://www.shimadzu.com/
Shimadzu is one of the world's largest manufacturers of analytical instrumentation, supporting applications in a broad range of industries, including life sciences, pharmaceuticals, food safety, environmental, cannabis, clinical research, and forensics. Delivering precise, reliable results, our robust platforms, including chromatography, multiplexing and mass spectrometry instruments, will improve productivity and efficiency in your laboratory. Visit our booth to learn about these efforts along with new Shimadzu platforms, including the Nexera QX multiplexing system, ultra-fast LCMS-8060NX triple quadrupole MS, CLAM-2030 clinical research lab MS automation platform, benchtop MALDI-8030, and automated protein digestion workstations.

Tecan (Booth #5)
http://www.tecan.com
Tecan is a leading global provider of laboratory instruments and solutions in biopharmaceuticals, forensics, and clinical diagnostics for pharmaceutical and biotechnology companies, diagnostic laboratories and design houses. Our suite of solid phase extraction and the innovative NBE (Narrow Bore Extraction) Format utilizes micro particulate sorbent allowing for greater contact of the analyte with the resin. This enables sample matrices to be more effectively removed and reduce interfering peaks in LCMS analysis. The NBE (Narrow Bore Extraction) format offers a redesign of conventional SPE columns to deliver substantial workflow improvement and automation friendly designs. This allows users to easily upgrade at any time when using NBE format to a fully automated workflow with Tecan's automation solution.

Thermo Fisher Scientific (Booth #16-18)
http://www.thermoscientific.com
From proven clinical laboratory services and diagnostics to scalable translational research solutions, we are a partner you can trust who will help you efficiently develop and apply clinical applications today, and for many years into the future. Our portfolio of Chromatography and Mass Spectrometry solutions are designed to empower clinical laboratories around the world with flexible research to lab developed test solutions and fully automated diagnostic testing capabilities. Our product portfolio offers a full range of benefits from clinical analyzer systems with complete assay kits, to scalable systems providing you flexibility in providing laboratory developed testing services.

UTAK Laboratories (Booth #8)
http://www.utak.com
At UTAK, we're proud to call ourselves "control freaks", but not in the way you might think. That's because our obsession lies not in taking control but in giving control—to the testing labs that need the finest quality

control materials for their clinical and forensic toxicology test methods. Our close-knit group crafts the quality controls these labs depend upon for every kind of analysis, including a wide range of comprehensive stock controls in urine, serum, blood, oral fluid and more, as well as starting matrices for laboratories seeking to develop in-house quality control material. We also create personalized control solutions to support the new methods these labs develop. Our dedication is grounded in our belief that better control for testing labs leads to more accurate results and ultimately, to better safeguarding of health and safety standards.

Waters (Booth #6-7)

https://www.waters.com/waters/global/Clinical-Diagnostics/nav.htm?cid=10009164

Waters is an innovator in chromatography, mass spectrometry, and thermal analysis instruments and software serving the life sciences industry. We deliver scientific insights to improve human health and well-being. As a clinical diagnostic laboratory, quickly providing the broadest range of tests with the best accuracy and selectivity is key, but many challenges can stand in the way. Current techniques can suffer from cross-reactivity that skew results, availability of reagent kits for new tests can be slow, and sample preparation is often too manual. Waters unlocks the potential of science by helping labs provide better care. Waters helps our customers enhance workflow efficiencies, improve specificity of results, and implement new tests with LC-MS/MS solutions, helping you provide accurate diagnosis and treatment without compromising compliance, flexibility, and performance.

YYZ Pharmatech (Mini-Table)

https://yyzpharmatech.com/

YYZ Pharmatech endeavors to enable the next generation of diagnostic & therapeutic medicine by quantitatively describing biology with greater sensitivity than ever before. YYZ is keen to collaborate with industry partners to address unmet clinical needs, where discovery & measurement approaches are currently limited by sensitivity. Technology Platform Highlights: Discover with DPiPSA™ (Database-Driven Proteome Intact Partition Statistical Analysis) - a proprietary bioinformatics platform that generates large scale blood peptide and protein databases, enabling a new paradigm of novel biomarker and target discovery. Measure with ELiMSA™ (Enzyme Linked Mass Spectrometric Assay) - a patented detection method with industry-leading sensitivity, quantification and reliability to identify any biomolecule, using a combination of enzyme amplification with detection by mass spectrometry.

Zef Scientific (Booth #27)

http://www.zefsci.com

•Is your Mass Spectrometer showing the uptime that you expect? •Do the different vendors tend to blame each other—or your method—for an issue? •Are you looking for a more harmonized and seamless experience in maintaining your LC-MS/MS? ZefSci is the country's premier independent LC-MS/MS engineering firm. A network of experienced field service and qualification engineers are strategically positioned nationwide supplying our customers with the highest level of services on AB/Sciex, Thermo, Waters, Agilent, and Shimadzu. 1- Service Contracts, 2- Preventative Maintenance, 3- Repair, 4- GxP Compliance IQ/OQ/PQ.

Zivak Technologies (Booth #34)

http://www.zivak.com

Based in Florida, Zivak Technologies is an international company providing calibrators and controls for a variety of analyses in the clinical diagnostic field. The company also offers its own fully automated UHPLC system FX-2000 with built-in sample preparation and injection, which enables laboratories around the globe to turn their LC-MS/MS instruments into a walk away tool for routine analysis by full automatization.

Presenter Index